HEALTH FOR THE PACIFIC 7

Body Systems

in Papua New Guinea

Richard Jones and Jennifer Miller

OXFORD
UNIVERSITY PRESS
AUSTRALIA & NEW ZEALAND

Oxford University Press is a department of the University of Oxford.
It furthers the University's objective of excellence in research,
scholarship, and education by publishing worldwide. Oxford is a registered
trademark of Oxford University Press in the UK and in certain other countries.

Published in Australia by
Oxford University Press
Level 8, 737 Bourke Street, Docklands, Victoria 3008, Australia

First published 2011
Reprinted 2023(D)

ISBN 978 0 19 557597 2

Typeset by Leigh Ashforth
Illustrated by Dimitrios Prokopis, Paul Konye, Rob Mancini and Uramina & Nelson
Proofread by Emma Short
Printed and bound in Australia by Ligare Book Printers Pty Ltd

The author and the publisher wish to thank the following copyright holders for reproduction of their material.

Getty Images/Dan McCoy – Rainbow, p.22; Lochman Transparencies/Alex Steffe, p.14

Every effort has been made to trace the original source of copyright material contained in this book. The publisher will be pleased to hear from copyright holders to rectify any errors or omissions.

Contents

Foreword

The *Health for the Pacific* series is intended to teach young people about important health issues that are affecting the lives of young men and women in the Pacific.

It is important that children and young people understand how their body works, grows and changes so they can learn how to keep themselves healthy and active. Young people are very curious about their bodies and how they work. This book aims to give a simple, accurate and interesting introduction to the systems of the human body.

Many health problems are linked to a lack of care or understanding about the body. The body is a complex machine, and good health depends on knowing how it works. Learning about the body and its organs, senses and systems is an important part of a young person's education.

Acknowledgments

The text is written to support the teaching of Personal Development in Papua New Guinea primary schools and secondary schools from Grade 6 onwards. It is a textbook for students and a resource book for teachers.

We would like to thank the many dedicated teachers and health workers who teach about the body and how to look after it, and who help guide young people through the challenges in their life. We were taught to be curious about and fascinated by the human body by our teachers. We dedicate this book to them.

Richard Jones and Jennifer Miller

 Notes for teachers

This textbook is written to be used by Upper Primary and Lower Secondary students and their teachers.

The knowledge, skills and attitudes in the text develop these learning outcomes from Personal Development:

Personal Development Grades 6–8

6.4.2 Explore influences of inherited characteristics and environmental factors on growth and development

7.4.1 Explore the functions of different systems and parts of the body

Personal Development Grades 9–10

9.2.1 Describe the major body systems and explain their functions during physical activity

10.1.1 Explain the functions of the male and female reproductive anatomy with respect to conception and pregnancy

There are many activities in the text for your students to complete and discuss. These can be used for self-study or as teaching and learning activities in class. They are designed for maximum student participation and developing life skills.

Introduction

The human body is an incredible machine. It moves, eats, grows, feels and repairs itself. Your body does most of these things all day, every day without you having to think about it. We need to keep our bodies healthy or we will become sick and die. If you look after your body and treat it with respect it will last for many, many years.

Your own body is unique to you but all human beings have the same **body systems**:

- senses and brain
- muscles, joints and bones
- heart and blood
- lungs and breathing
- stomach and digestive system
- reproductive system (a different one for males and females)
- protection and healing system.

If you know how your body works you can look after it better. This book will take you on a tour of your body systems.

Chapter 1 What am I made of?

We are animals

All animals share the same features. All animals:

- move
- use oxygen and release carbon dioxide
- eat and digest food
- fight off infections
- grow
- reproduce
- sense the world around them.

Human beings are animals, too.

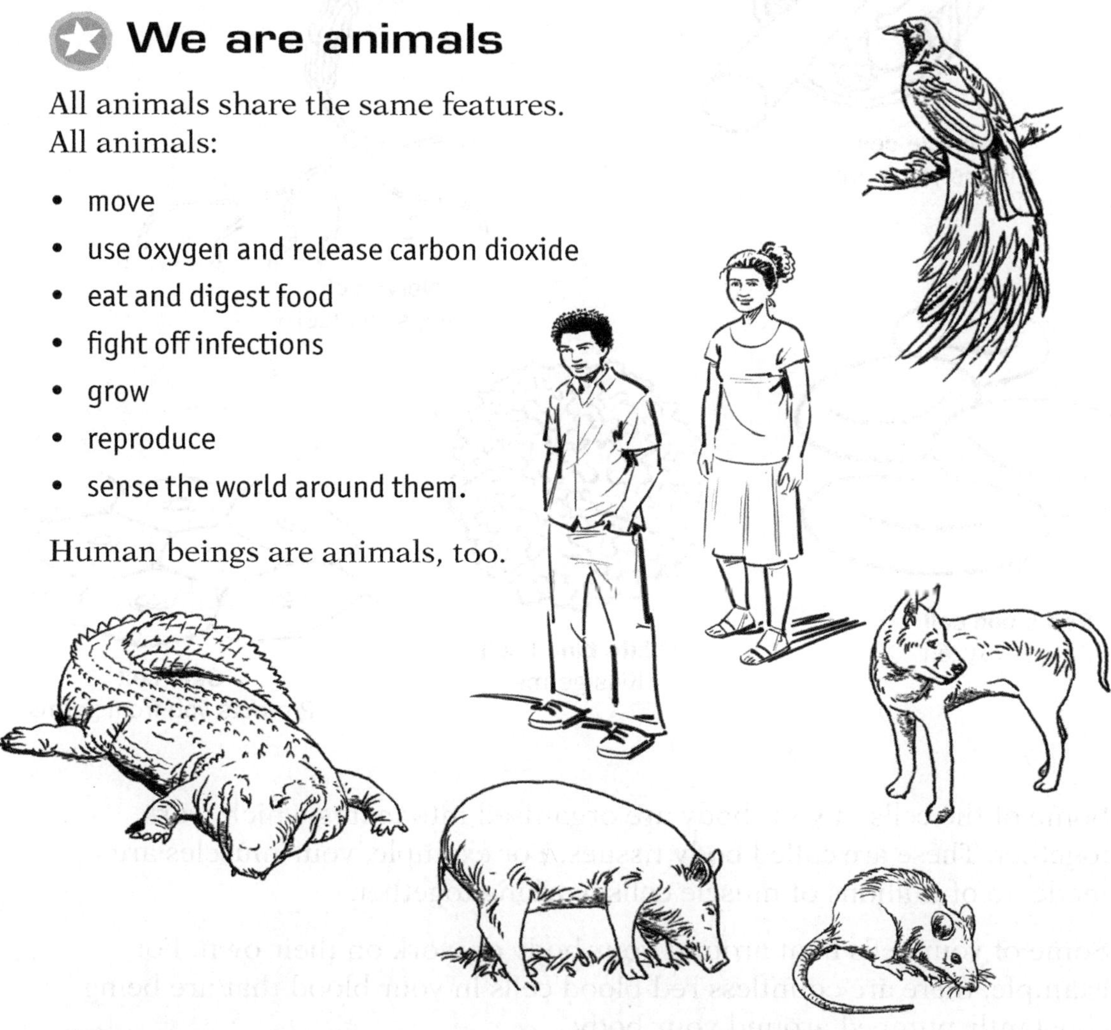

What makes a human being?

Every human being, including you, is made up of millions and millions of tiny **cells**. You can't see cells without looking at them through a microscope. There are many different types of cells and each type has a specific job to do in your body.

Some of the millions of cells that make up you!

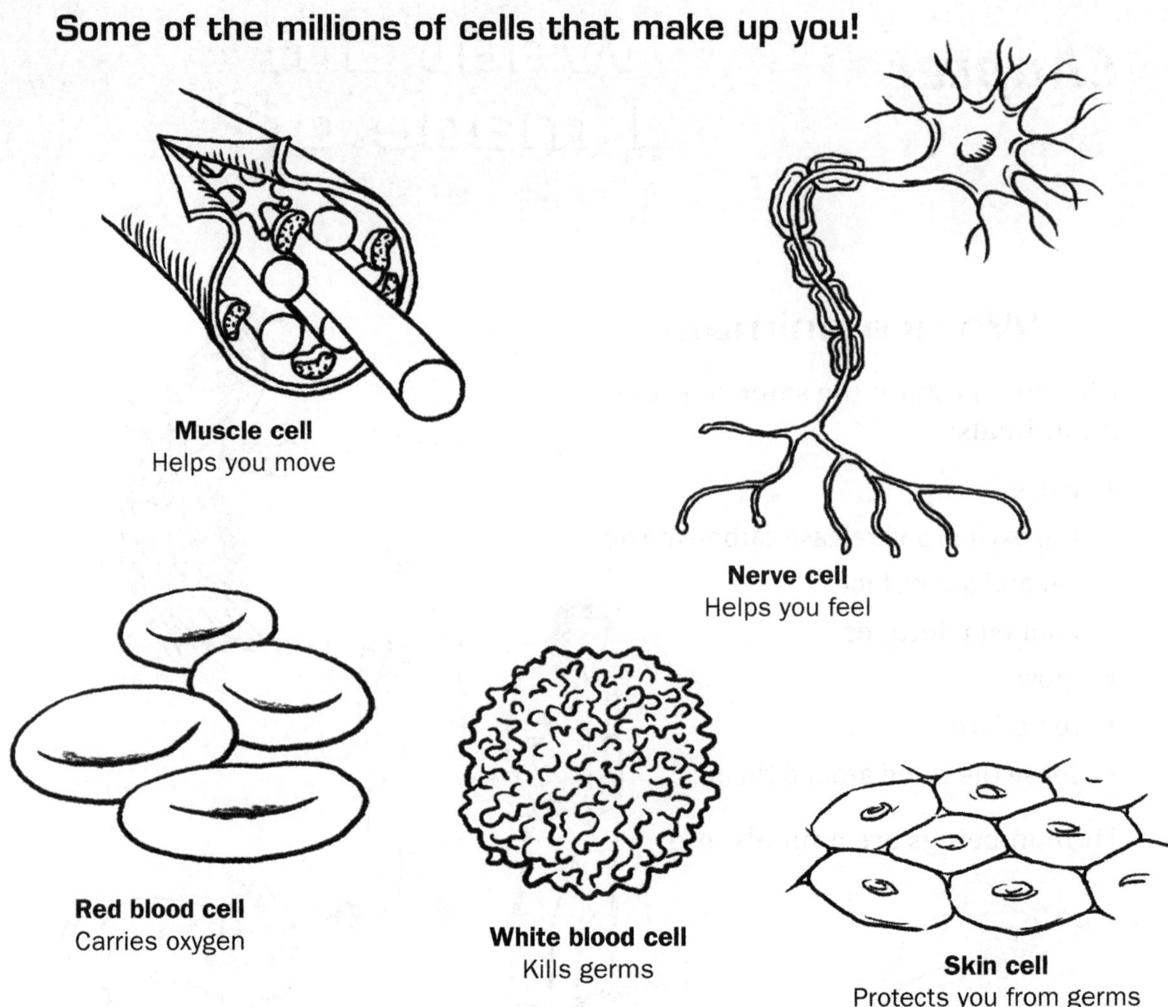

Some of the cells in your body are organised into teams which work together. These are called body tissues. For example, your muscles are made up of millions of muscle cells working together.

Some of your cells float around your body or work on their own. For example, there are countless red blood cells in your blood that are being constantly pumped around your body.

Cells are always being made and destroyed. Some cells, like bone cells, can stay alive for many years, but some last only a few days. Your skin cells, for example, are always being replaced. Most of the dust on your pillow and sheets is dead skin cells!

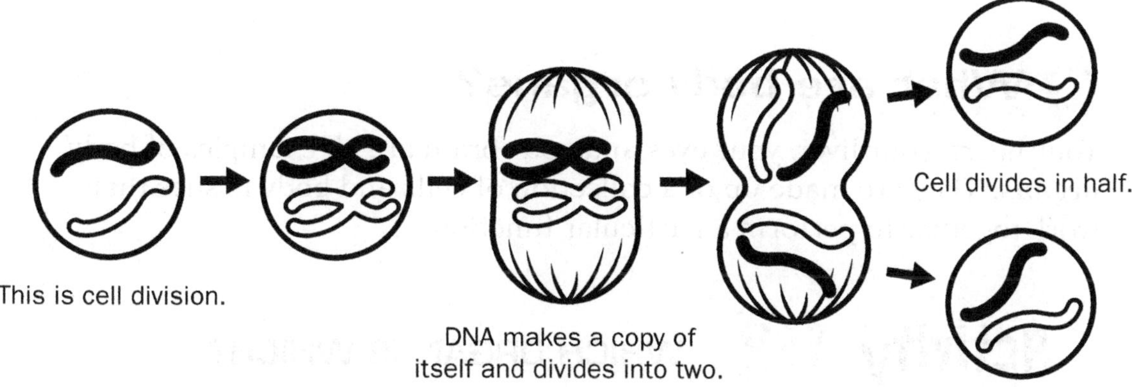

Most cells are made when a cell divides in half. Before you were born, you were just one cell. Now you are made up of billions of cells.

Almost every cell in your body contains **DNA**. DNA contains the instructions that control the development and function of all cells. The parts of DNA that carry specific information are called **genes**. Every human being has a different set of DNA, which is why we all look different. Every person gets half of their DNA from their mother and half of their DNA from their father. This is why we sometimes look a little like our parents or grandparents.

What makes us human?

Every living thing has DNA in its cells. Many parts of our DNA are the same as the DNA in other living creatures. In fact, 96 to 98% of our DNA is the same as the DNA of chimpanzees, who are our closest living relative.

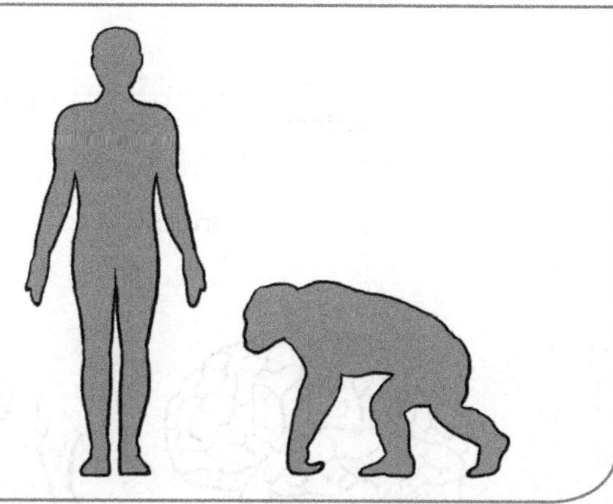

What is special about identical twins?

Identical twins are special and rare. They both have the same DNA, so they look exactly the same!

This is because they grew from one cell that split apart before they were born.

What are body organs?

Your heart, your liver, your eyes and your brain are all examples of body **organs**. They are made up of a collection of cells and body tissues that work together to perform a particular function.

Activity 1·2 WHICH ORGAN IS WHICH?

Match the organ with the correct name.

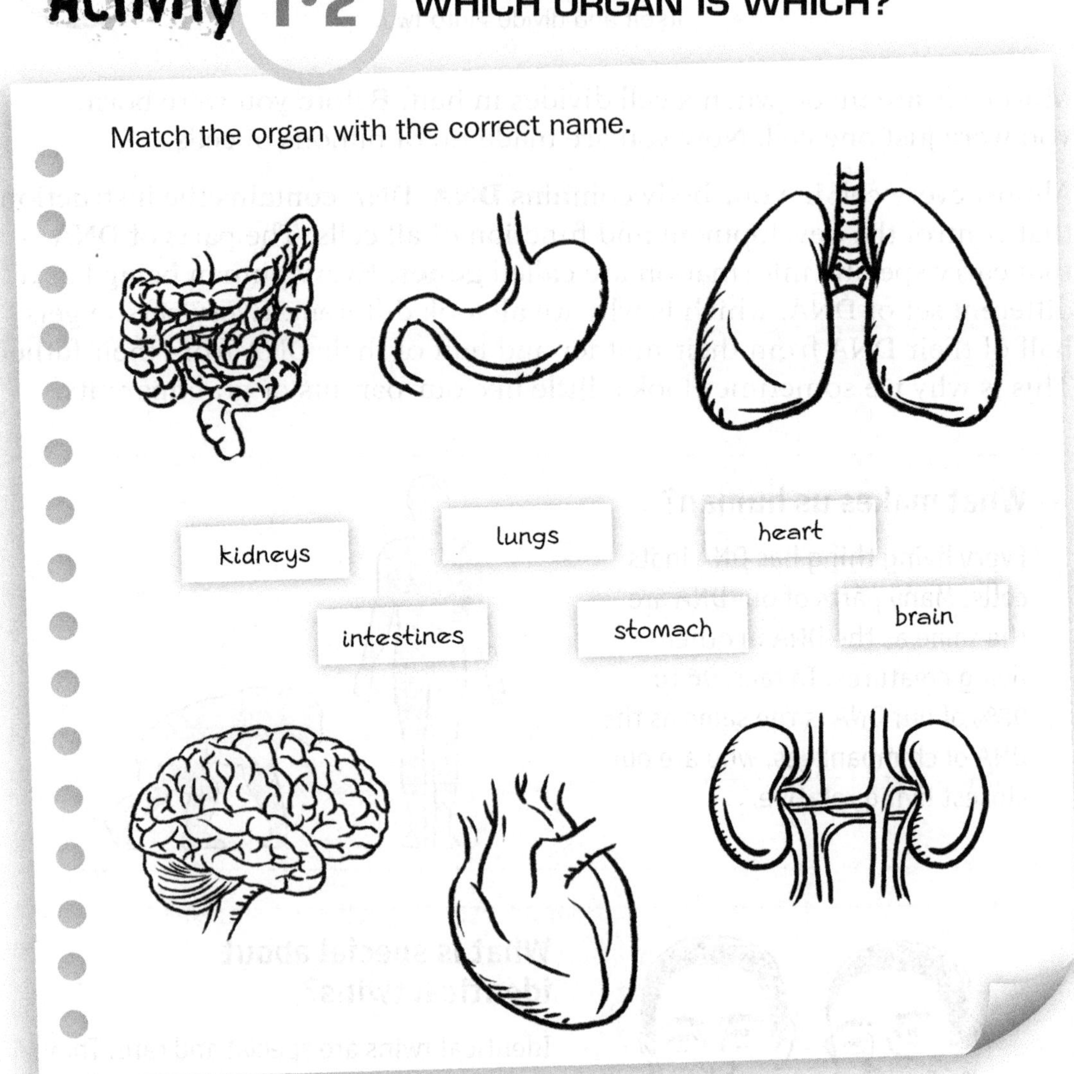

Body organs have special jobs to do in your body. Your heart is a pump which sends blood around your tissues. Your liver cleans up waste. Your lungs take oxygen from the air to keep you alive.

What are body systems?

Many organs and body tissues work together in teams. These are called body systems. There are many different body systems in your body.

Each of the remaining chapters presents a different body system, what it does and how it works.

Body system	What the body system does	The organs and body tissues in the body system	Where to read about the body system
Musculoskeletal system	Movement and protection	Bones and muscles	Chapter 2: Muscles and bones
Circulatory system	Distributes fuel around the body	Heart, veins and arteries	Chapter 3: Heart and blood
Respiratory system	Breathes in oxygen and breathes out carbon dioxide	Lungs and chest muscles	Chapter 4: Breathing
Nervous system	Senses the world, sends information and thinks	Brain, nerves, skin, eyes, ears, tongue	Chapter 5: Nervous system
Reproductive system	Makes new human beings	Females: ovaries, oviducts, uterus, vagina Males: testes, penis	Chapter 6: Reproductive system
Digestive system	Extracts fuel from our food	Teeth, stomach and intestines	Chapter 7: Digestive system
Excretory system	Removes waste	Liver, kidneys, intestines and bladder	Chapter 8: Excretory system
Immune system	Protects the body and fights germs	Lymph nodes	Chapter 9: Immune system
Endocrine system	Communicates within the body using hormones	Hormones, endocrine glands, adrenal glands	Chapter 10: Messaging system

Activity 1·3 SORTING AND RANKING

Which body system would you like to learn about first?

1. Working in a small group, write the names of each body system on separate pieces of paper.

digestive system | reproductive system | circulatory system

nervous system | musculoskeletal system | excretory system

immune system | endocrine system | respiratory system

2. Discuss which body system is the most important and which one you would like to learn about first. Think about the best order to learn the systems and explain your reasons.
3. Now arrange the pieces of paper into a pyramid. At the top of the pyramid will be the body system your group thinks is the most important. This will be the first chapter you read.
4. Place the next two most important body systems under the first piece of paper.

Most important (will study first)

5. In the next row, place the three body systems you will look at next.
6. Finally, rank the remaining body systems. This is the order in which your group wants to work through the book. Now find the right chapter for the most important body system and start studying.

1st to read

2nd | 3rd

4th | 5th | 6th

Chapter 2 Muscles and bones

Your bones and muscles support, protect and help you move. Without them you would flop over like a slug or a jellyfish! The system of muscles and bones is called the musculoskeletal system.

 What are bones?

Your **skeleton** is made up of 206 bones. These bones are alive and they are incredibly strong and surprisingly light.

Bones are different sizes. The largest bone is your femur (thigh bone). The smallest is a tiny bone in your ear.

Your bones are built with a very hard and strong outer layer and a soft spongy inner part. Your bone cells make a special honeycomb structure called **bone marrow** which gives the bones their amazing strength.

Different bones perform different functions in your body.

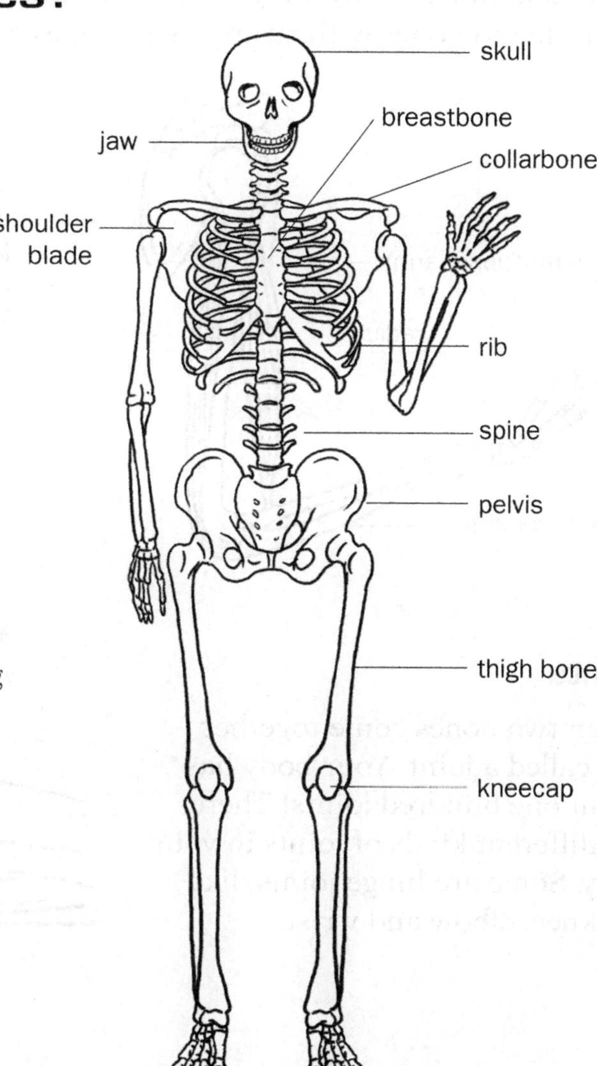

Body armour bones

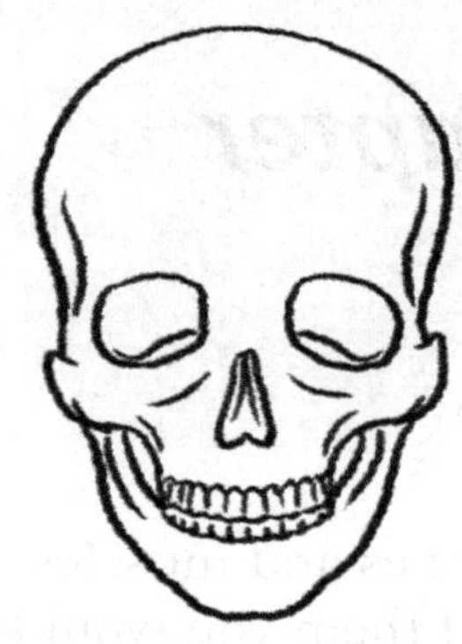

Some bones protect your soft organs and body tissues. There are over twenty bones in your skull and face. These bones protect your brain, mouth and eyes. Your ribcage bones protect your heart and lungs. Without the protection and support we get from these bones we would get injured very easily.

Bones for moving

Some bones help us to move. These bones have muscles attached. Bones and muscles work together to allow us to move. Muscles are connected to bones with strong living ropes called tendons.

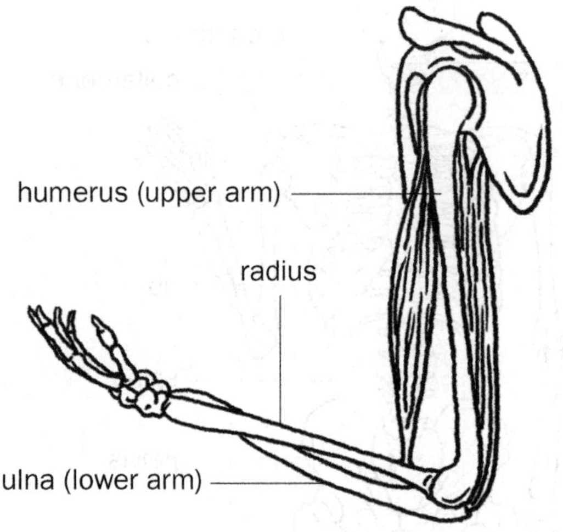

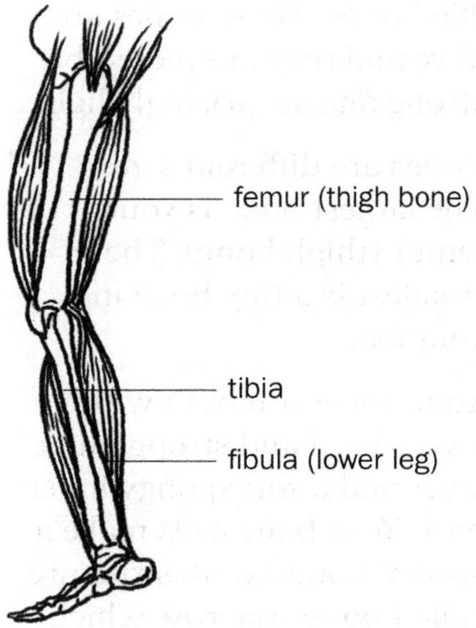

Joints

When two bones come together it is called a joint. Your body has about one hundred joints! There are different kinds of joints in your body. Some are hinge joints, like the knee, elbow and wrist.

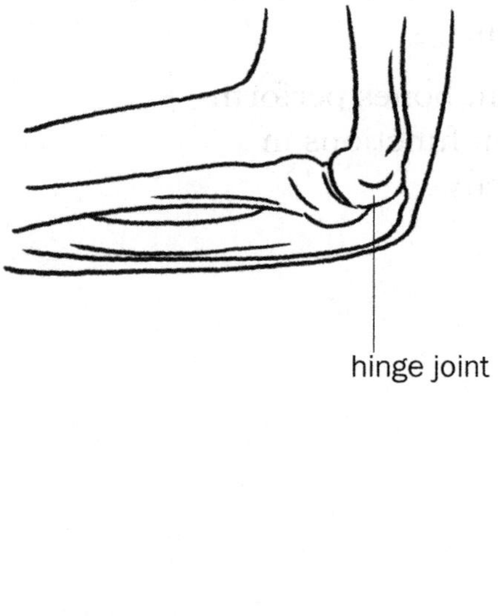

Some joints are special ball-and-socket joints, like the shoulder or hip, which allow us to move our arms and legs in many different directions.

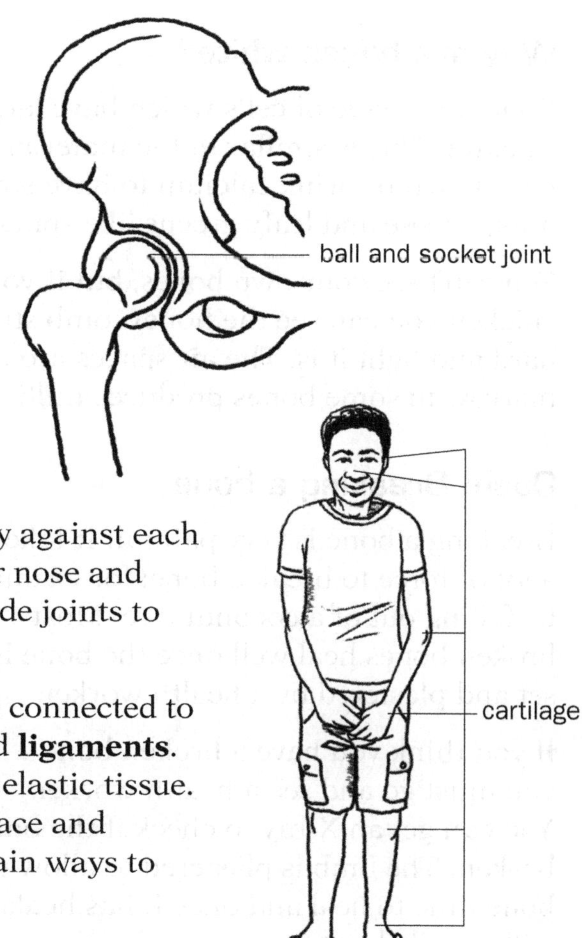

Bones are connected to one another in different ways. Some bones are connected to one another by **cartilage**, which is a rubbery material. This is for joints where only a little bit of movement is allowed. Cartilage stops the ends of bones from rubbing directly against each other. You can feel cartilage in your nose and ears, but it is also hidden away inside joints to stop the bones rubbing together.

At most of our joints, the bones are connected to each other with strong cables called **ligaments**. These are small and thick bands of elastic tissue. Ligaments help to keep bones in place and prevent bones from moving in certain ways to protect them from breaking.

Activity 2·1 WHICH BONES DO WHAT?

Carry out an investigation on your own skeleton.

1 How many hinge joints can you find on your own body?
2 How many ball-and-socket joints can you find?
3 Try to count all your rib bones.
4 Which is the longest bone you can find on your body? How long is it?
5 Which is the shortest bone you can find on your body? How long is it?

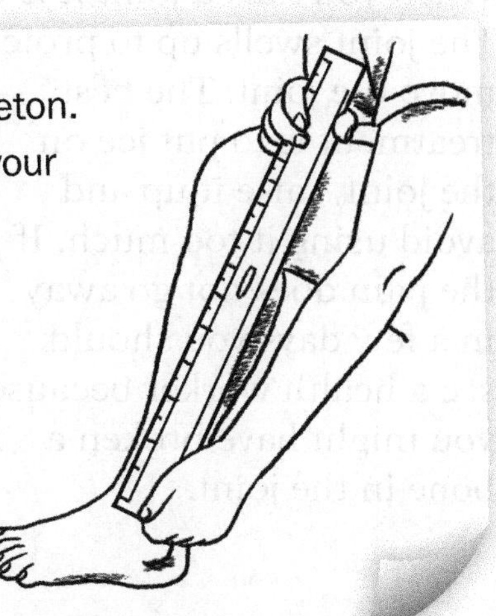

Why are bones white?

Bones are made of cells which have large stores of a mineral called calcium. This is similar to the material found in sea shells. Human beings need to eat or drink calcium to have strong bones. You can find calcium in milk, cheese and leafy greens like spinach.

You can't see your own bones, but if you look at a bone from a pig or a chicken you can see the honeycomb structure in the bone and feel how hard and light it is. The air spaces are part of the bone marrow. The bone marrow in some bones produces millions of red blood cells every second.

Ouch! Breaking a bone

Breaking a bone is very painful! It takes a lot of force to break a bone; for example, by falling out of a coconut tree. Most broken bones heal well once the bone is set and plastered by a health worker.

If you think you have a broken bone you must go and see a health worker. You can get an X-ray to check if the bone is broken. The limb is plastered to allow the bone time to heal and once it has healed it will usually be stronger than before.

Ouch! Hurting a joint

When you hurt a joint it is usually the ligaments that have been damaged. The joint swells up to protect the damaged tissues and it is painful to move the joint. The best treatment is to put ice on the joint, raise it up and avoid using it too much. If the pain does not go away in a few days you should see a health worker because you might have broken a bone in the joint.

Activity 2·2 BROKEN BONE SURVEY

Find someone in your school, family or village who has had a broken bone.

1 Which bone did the person break?
2 How did they break it?
3 What did they do about it? How was it treated?
4 How long did it take to heal?
5 Does the person have any problems with the bone now?

What are muscles?

Without your muscles you would not be able to move. Muscles are made up of long fibres which are made up of millions of muscle cells. They are well supplied with food and oxygen because muscles work very hard. There are so many muscles in your body that they make up about a third of your body weight!

Some muscles need you to control them with your brain. These are called **voluntary muscles** and include the muscles in your arms and legs. But there are some muscles which work without you having to think about them. These are called **involuntary muscles** and they work even when you are asleep or doing something else. Blinking, breathing and your heartbeat are all caused by involuntary muscles.

How do my muscles work?

Your muscles have many different nerve cells around them. These nerve cells connect your muscles to your brain. When your nervous system sends an electrical signal to your muscles they contract (become shorter). When the muscle becomes shorter, it pulls the bone it is attached to and causes it to move. Flex your arm and you can see your muscles work. Muscles often work in pairs so the joint can be pulled back to its original position by an opposing muscle.

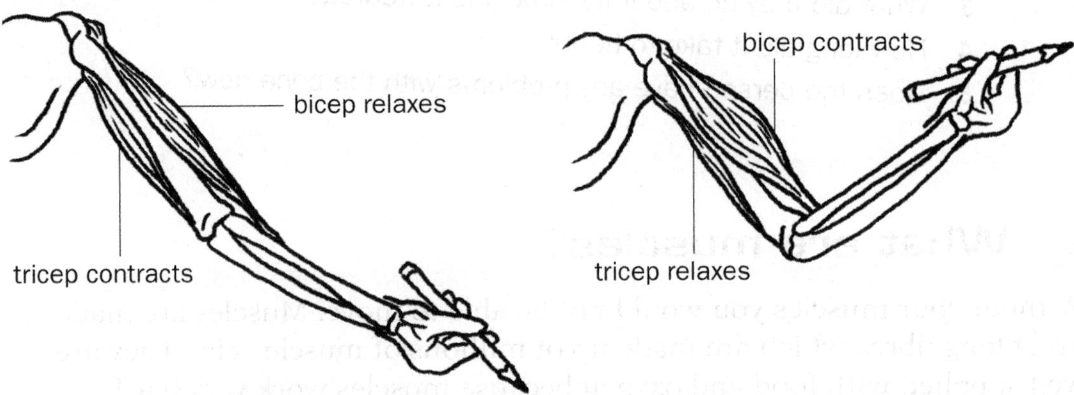

Muscles are attached to your bones by **tendons**. You can feel these tendons under your skin at different parts of your body. They feel like thick, strong, rubbery strings. You can feel your tendons in your wrists, the back of your hands, the back of your ankles and the bottom of your neck.

Ouch! Damaging a muscle

Most of us have hurt a muscle at some time.

Cramp

Cramp occurs when the muscle becomes exhausted and tenses up. You can treat a cramp by stretching the muscle, drinking water or rehydration fluids and resting.

Sore muscles

If you work your muscles too hard or use them too much you will feel very sore the next day, especially if you have not exercised that muscle much before. You can treat sore muscles by stretching before and after exercise, massaging the sore area, and putting ice on the muscles.

How to look after your muscles

You can make your muscles stronger and fitter by exercising. If muscles work hard regularly they become stronger. Playing sport, walking, running, swimming and dancing are all great for muscles. Make sure you eat and drink enough energy foods and water.

Most muscles get tired out if you exercise too much, but one muscle is different. This is your heart. It never gets tired out!

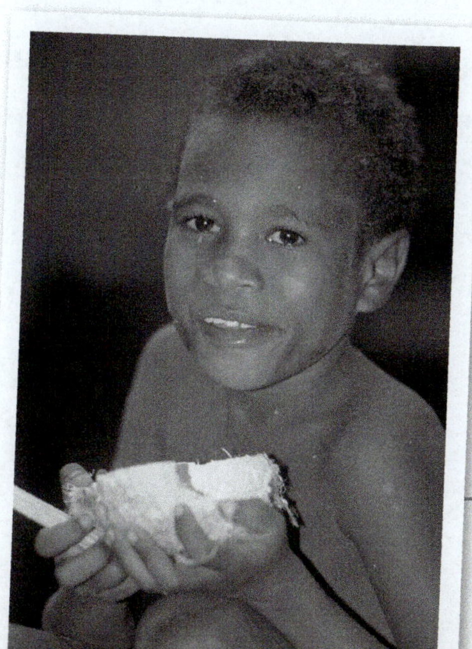

Amazing muscle fact

Where do you think the strongest muscle in your body is? Your arm? Your leg?

The strongest muscle is in your mouth!

Activity 2·3 STRENGTH TEST

How strong are your muscles? Design investigations to see how strong your muscles are and then use the results to compare your muscles to your friends' muscles.

Design a fair test for:

1. your finger muscle
2. your arm
3. your leg.

Remember: if you exercise a muscle, it becomes stronger!

Chapter 3 Heart and blood

Your circulatory system is made up of your blood, heart, **arteries**, capillaries and **veins**. The job of this body system is to carry your blood throughout your body to supply every cell with fuel and oxygen and collect waste. If your cells do not get food and oxygen they will die. The circulatory system also helps the body to fight diseases and to keep your body's temperature at a constant level.

Heart

Your heart is one of the few organs you can actually feel working. When you stop after you have been running, for example, you can feel your heart's powerful beat. The heart is a special muscle inside your chest, protected by the ribs. Its job is to pump blood around your body.

The heart is so good at pumping blood that it only takes about a minute for a blood cell to travel around your entire body.

Your heart is actually made up of two pumps. Each pump is built with two chambers: the **atrium** and the **ventricle**. Blood enters the heart through the atrium and leaves the heart through the ventricle. One side of the heart pumps blood to the lungs and back to the heart. The other side pumps the oxygen-rich blood from the heart to the rest of the body.

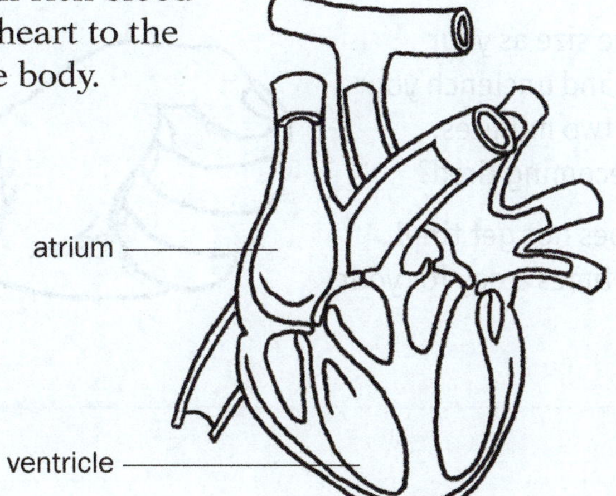

Special valves keep the blood moving in the right direction. When a doctor listens to your heartbeat through a stethoscope, he or she is listening to the heart valves opening and closing.

Every time the heart contracts it sends a pulse of blood to the body. You can feel this **pulse** by pressing your fingers against your neck or wrist.

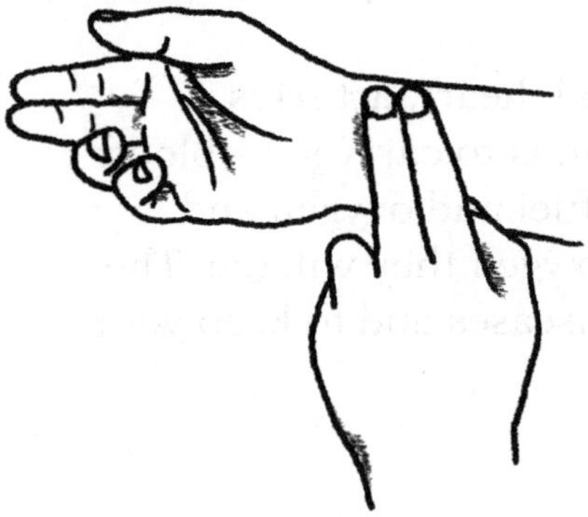

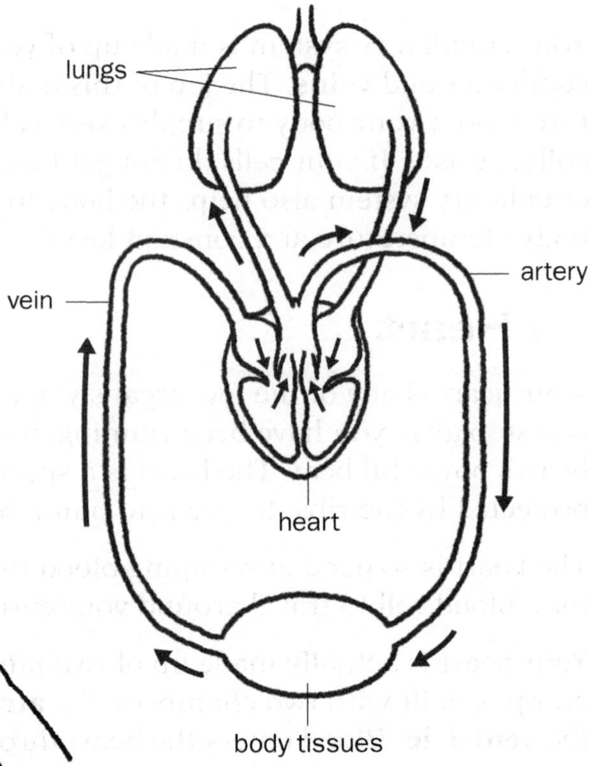

lungs
artery
vein
heart
body tissues

Amazing heart fact!

Your heart is about the same size as your clenched fist. Try to clench and unclench your fist over and over again for two minutes. Do you feel your muscles becoming tired?

The special heart muscle does not get tired. It will pump about 100 000 times a day for your whole life!

Exercising

When you run or jump or swim, the cells in your muscles need more oxygen, so your heart beats faster and harder to move the blood to the lungs and the tissues more quickly. The harder you exercise, the harder your heart beats until it reaches a maximum speed.

Activity 3.1 MEASURING YOUR HEART RATE

How fast will your heart beat? How quickly does it speed up and slow down?

1. Measure your heart rate when you are resting by recording how many times your heart beats in one minute (beats per minute or bpm).
2. Run on the spot for one minute and measure your heart rate again.
3. Then run on the spot for two minutes and measure your heart rate again.
4. Finally, rest for one minute and measure your heart rate again.
5. What do you notice? How does your bpm change? How else does your body change?
6. Record your results in a table or graph.

Activity 3.2 HEART RATES

Look at these heart rates of different animals in beats per minute. What patterns do you notice?

Resting heart rates usually depend on the size of the animal. The smaller the animal, the faster the heart rate. However, the fitter you are, the slower your resting heart rate.

Newborn baby	130 bpm
Baby	120 bpm
Five-year-old child	95 bpm
Ten-year-old child	80 bpm
Average woman resting	76 bpm
Average man resting	68 bpm
Mouse	600 bpm
Rabbit	200 bpm
Dog	125 bpm
Pig	80 bpm
Elephant	35 bpm

What are arteries?

The blood that is pumped from your heart throughout your body is under great pressure. It has to get to every cell in your body! This oxygen-rich blood is carried in special tubes called arteries which have thick muscular walls to stop them bursting. In a few places on your body, the arteries are close to the surface of the skin and you can feel the pulses of blood. In your neck you can feel your carotid artery, which sends oxygen-rich blood to your brain.

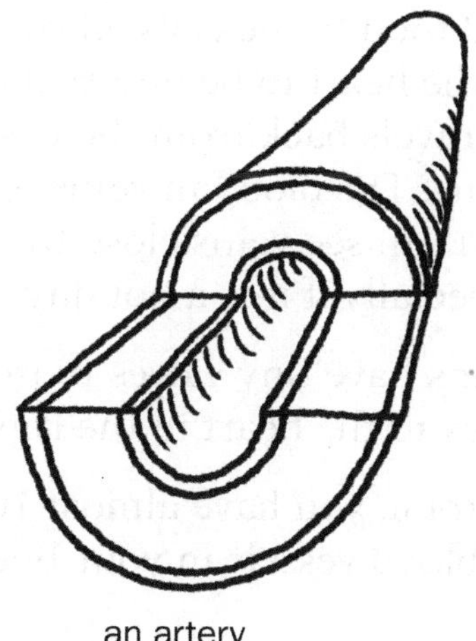

an artery

The blood is pumped from the heart to the arteries and then into the capillaries.

What are capillaries?

Capillaries are smaller tubes that carry the blood in a complicated network close to every cell. They are very thin so that food and oxygen can cross over to the cells.

You can see capillaries by looking at someone's eyeball. They are tiny little tubes full of red blood.

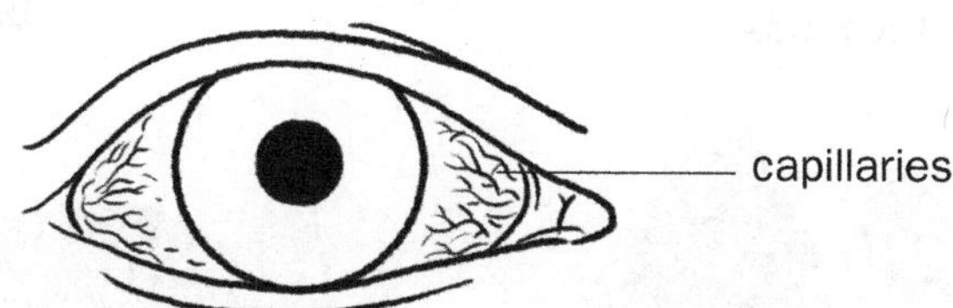

If you look at a white person on a hot day you will see their face going red. This is because the capillaries in the skin carry blood close to the surface to try to cool the person down. You can't see this happening in Melanesian people because they have dark skin.

What are veins?

After the blood has given up its precious cargo of oxygen and food to the cells all over the body, it has to travel back to the heart to be pumped to the lungs for fresh oxygen. It travels back from the tissues in soft tubes called veins. The blood in veins is not under a lot of pressure. You can see veins close to the surface of your skin, especially if it is a hot day or you are exercising.

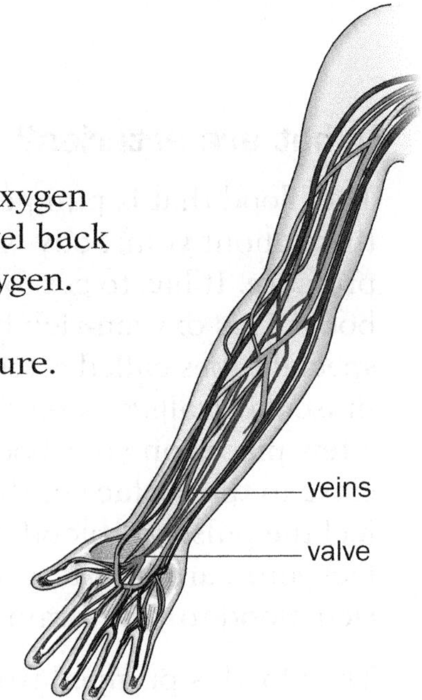

Veins have tiny valves to make sure the blood goes back to the heart in the right direction.

In total, you have almost 100 000 kilometres of blood vessels in your body!

Looking after your heart and arteries

It is very important to look after your heart and the rest of your circulatory system. You can do this by exercising, playing sport and eating a healthy diet of lean meat, fresh vegetables and fruits. You should avoid eating too many fatty and sugary foods like cakes, lamb flaps and fried food.

Some people in PNG do not eat properly and do not exercise enough.

They are at risk of heart disease.

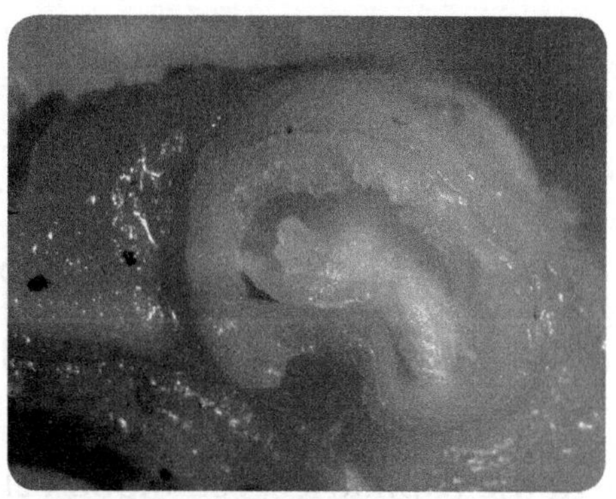

Overweight people have lots of extra fat. What you cannot see is the fat starting to block up the arteries that supply their brain and heart. If these arteries become blocked, the person could suffer a **heart attack** or a **stroke**. These can kill. If your heart stops beating then you will die

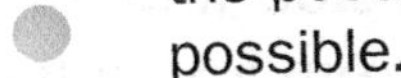

after about five minutes.

HEALTHY HEARTS

Design a poster for your local trade store or church that explains why it is important to eat healthy food and exercise more. Use your local language (*tok ples*) and select a picture and slogan that will make the poster as effective as possible.

What is in your blood?

Your blood is a very interesting and important liquid. Most adults have about four to five litres of blood moving throughout their bodies.

Blood is made up of several different parts:

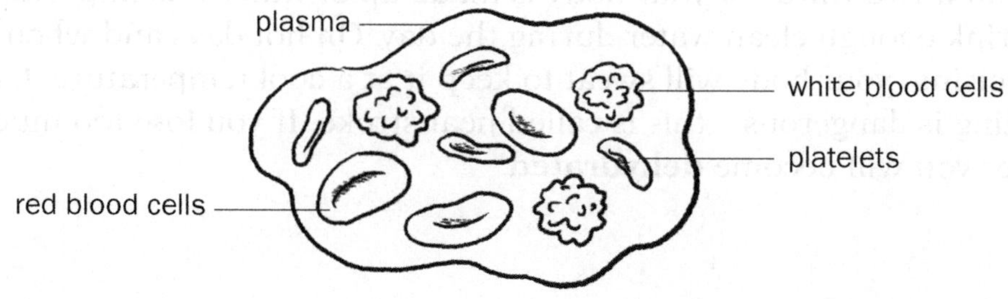

Blood has different jobs. It carries oxygen, food and water to the cells around your body. It then picks up waste carbon dioxide produced by the cells and carries it to your lungs. Blood also carries **white blood cells** which provide the body's defence against germs.

What are red blood cells?

Blood is red because of the **red blood cells**. They have only one job, which is to collect oxygen from the lungs and deliver it all around your body.

In the capillaries, red blood cells deliver oxygen to the tissues and collect waste carbon dioxide and carry it to the lungs. Carbon dioxide is a poison that your body needs to breathe out.

Red blood cells contain a special substance containing iron, called haemoglobin, which helps them to carry oxygen and carbon dioxide. Iron is found in red meat and green leafy vegetables.

You have more red blood cells in your body than any other kind of cell. The body is constantly making new red blood cells to replace old and damaged ones.

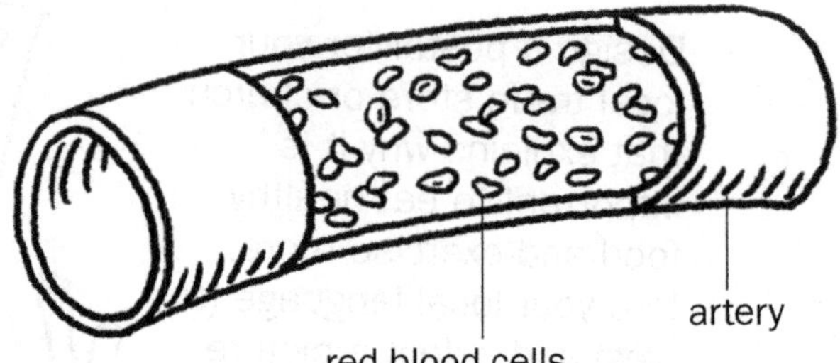

What are platelets?

When you have a cut, the blood flows for a while and then clots and forms a scab. This protects the wound as it heals and stops you from losing too much blood. The clot is made up of **platelets** in the blood, which stick together if you have a cut.

What is plasma?

Plasma is a watery liquid in which the cells in the blood float. It also carries water and food to the cells. When you have a blister, the liquid in the blister is mostly plasma and white blood cells.

As about two-thirds of your body is made up of water it is important to drink enough clean water during the day. On hot days and when exercising, your body will sweat to keep it at a cool temperature. Overheating is dangerous – this is called heat stroke. If you lose too much water you will become **dehydrated**.

Red blood cells and malaria

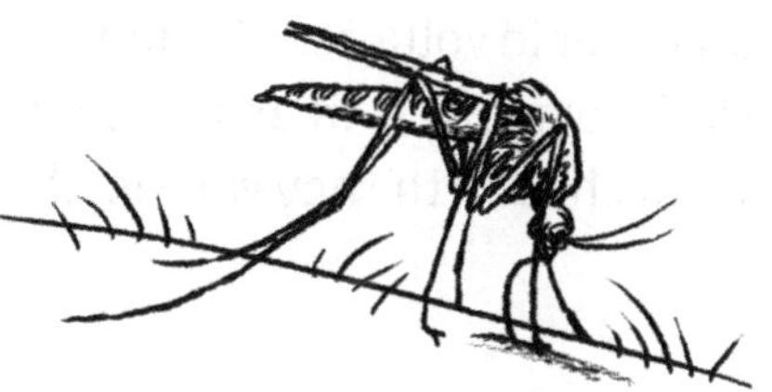

The malaria **parasite** lives part of its life cycle inside red blood cells. A malaria fever is caused when the parasites break out of the cells. This also kills the red blood cells, causing **anaemia**, which is very dangerous.

People who have had a malaria infection often need iron supplements or more iron in their diet to help replace red blood cells.

Activity 3.4 BLOOD ROLE PLAY

Work with a group of friends to prepare a role play which explains how red blood cells travel around the body delivering oxygen.

Think about how you will demonstrate:

- the pumping of the heart
- collecting oxygen in the lungs
- travelling quickly through the arteries
- delivering oxygen to the cells in the capillaries
- collecting waste carbon dioxide from the cells
- travelling slowly back to the heart through the veins
- doing it all over again and again!

Finally, show the effect of malaria on the red blood cells in your role play.

Present this role play to your class or village.

Giving blood, helping others

Many people volunteer to donate their blood to help other people who are sick. If someone is in an accident, has malaria or suffers from bleeding during childbirth they will need a **blood transfusion**.

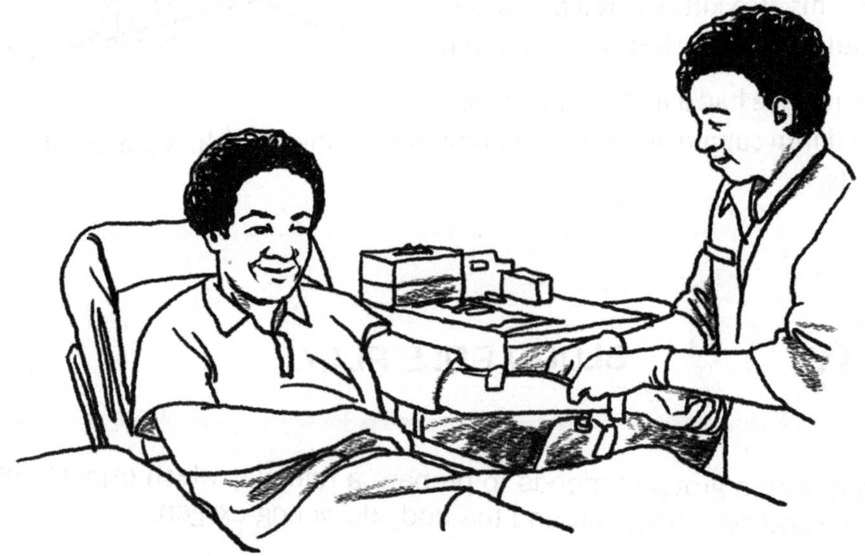

Before the blood is transfused into the patient, health workers will check the blood to make sure it is the right type and that the blood does not carry any diseases (like HIV).

The donor's body soon makes new blood cells to replace the blood they have donated. Blood donors save thousands of people every year. Do you know anyone who has donated blood? Will you be a blood donor one day?

Chapter 4 Breathing

Human beings are animals. We have to breathe in oxygen and breathe out carbon dioxide. If we stop breathing we will quickly die. The body system responsible for breathing is called the respiratory system.

All of the cells in your body need oxygen. When the cells respire they use up the oxygen and release carbon dioxide as waste gas. Both of these important gases are carried by red blood cells around the body through the circulatory system.

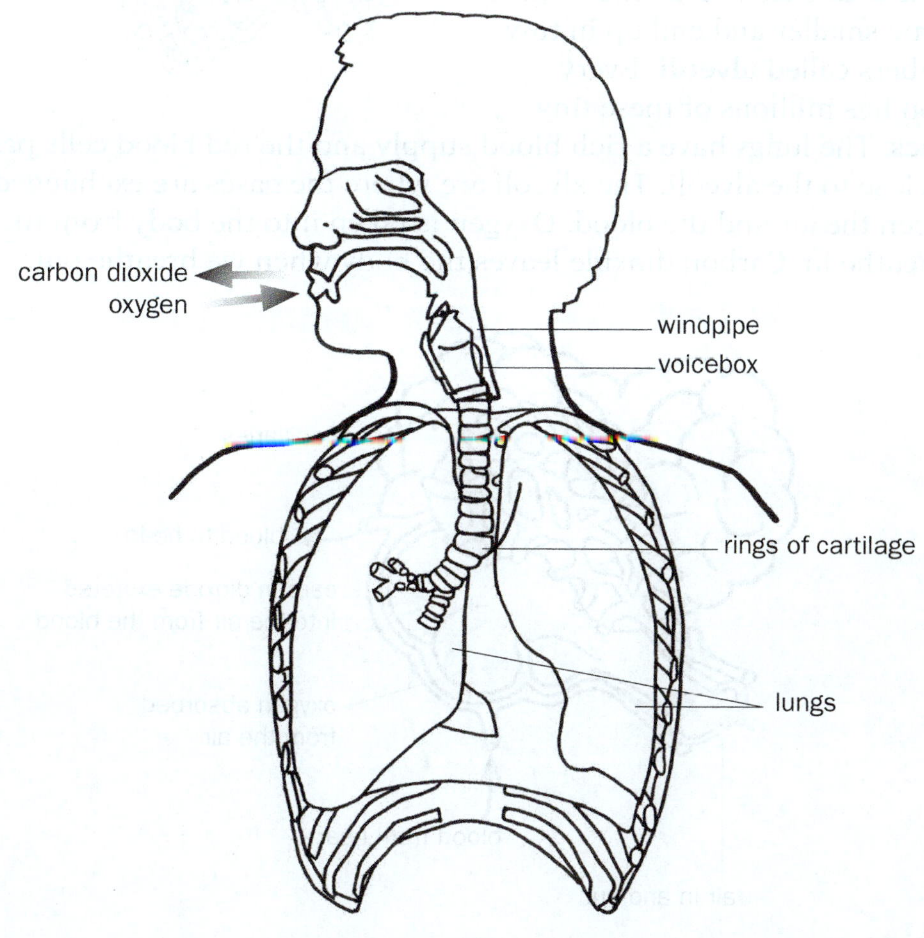

How do my lungs work?

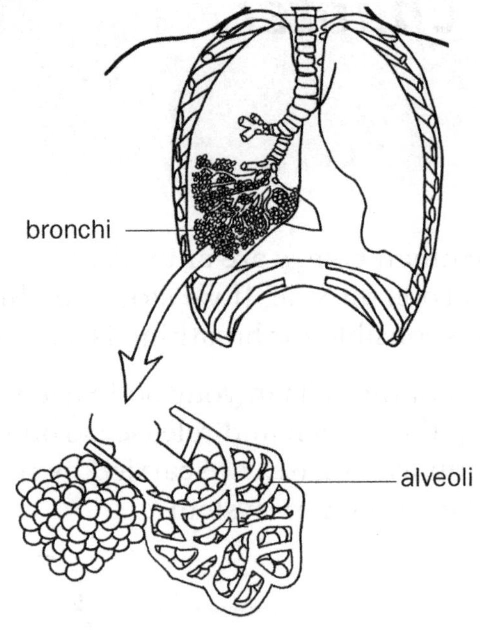

Your lungs are two soft and stretchy air bags full of millions of tiny tubes. Air passes down your windpipe (trachea) from your mouth and nose. The windpipe is a hollow tube protected by rings of cartilage (which you can feel with your fingers). On the way in and out of your lungs the air passes through your voice box (larynx).

Your windpipe splits into two smaller tubes and enters your two lungs. These tubes split again and again like the branches of a tree. The tubes become smaller and end up in tiny chambers called **alveoli**. Every person has millions of these tiny air sacs. The lungs have a rich blood supply and the red blood cells pass very close to the alveoli. The alveoli are where the gases are exchanged between the air and the blood. Oxygen is taken into the body from the air we breathe in. Carbon dioxide leaves the body when we breathe out.

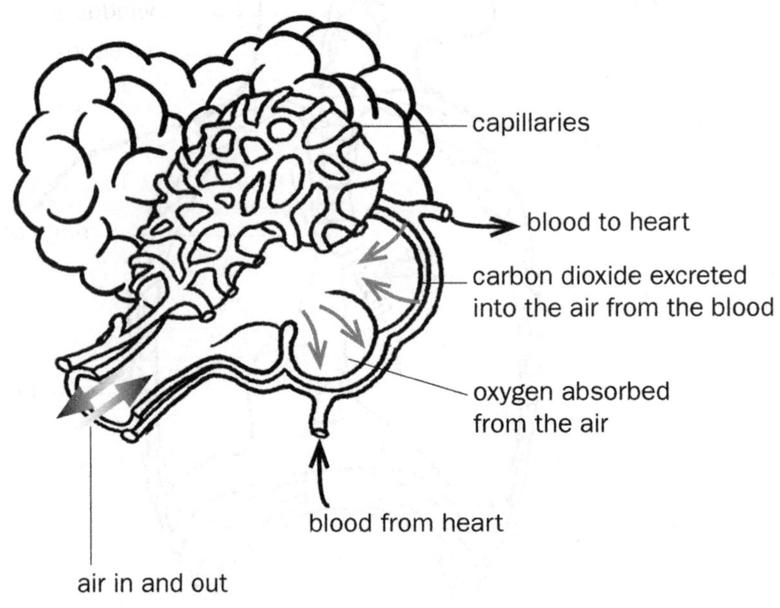

Moving the air in and out of the lungs is called breathing. Your body does this automatically most of the time. The lungs are protected by the ribcage, which can move in and out. Below the lungs is a strong sheet of muscle called the **diaphragm**. This pulls the lungs down to suck in air.

Stitched up!

Sometimes when you are exercising hard and get out of breath you feel a pain across your belly. This is called a stitch. It is your diaphragm becoming tired!

When you stop and get your breath back the stitch goes away.

Activity 4.1 HOW MUCH AIR DO YOUR LUNGS HOLD?

Work with a friend to investigate how much air your lungs can hold.

1 How will you safely measure how much gas your lungs can hold?
2 Does the size of the person's body make a difference to the size of their lungs?
3 Does the age of the person make a difference?

 Keeping your lungs healthy

It is very important to keep your lungs healthy and in good working order. The best way to do this is to exercise them. Playing sport, running, dancing and walking strengthens the muscles that help you breathe and increases your lung size. The harder your body works, the faster and deeper you breathe because you need more oxygen for your muscles.

However, there are some risks to your breathing, some of which you can avoid.

Smoking

Smoking cigarettes is very harmful. The poisons in the smoke damage different parts of your respiratory system.

Smoking damages the alveoli and fine tubes in the lungs, causing lung cancer and emphysema.

Smoking kills the fine hairs that keep the windpipe clean, leading to a smoker's cough.

There are poisonous gases like carbon monoxide in the cigarette smoke. This poison sticks to the red blood cells so they cannot carry enough oxygen to your body's tissues.

Activity 4.2 ROLE PLAY

1. With your friends, brainstorm the different situations where young people smoke.
2. How would you respond to peer pressure? With a friend, role-play answers to these situations.
3. Now write your role plays as a play script.

Lung infections

Most people suffer from lung infections at some point during their lives. These infections usually become better on their own but sometimes you need to take antibiotics. Some infections can be very serious.

Tuberculosis

Tuberculosis (TB) is caused by bacteria. It causes coughing and weight loss and attacks the body slowly. TB can be treated with a long course of medicine.

Pneumonia

Pneumonia can be caused by bacteria or a **virus**. It is very serious and can kill people very quickly. If someone is sick with pneumonia, especially a child or an old person, and they are struggling to breathe, they must be taken quickly to the health centre.

Asthma

Asthma is a kind of allergic reaction and is quite common in children. It is caused by a tightening of the tiny air tubes in the lungs. Sometimes this is caused by smoke or dust or getting out of breath. There are medicines available such as pumps or inhalers to help the person breathe easier.

Activity 4.3 BREATHING FOR LIFE!

Calculate how many times you breathe:

- in a minute
- in an hour
- in a day
- in a week
- in a month
- in a year
- in your life so far!

Yawning

Your body can detect changes in your blood. If your blood has too much carbon dioxide, your body will automatically make you breathe more deeply and more quickly.

Sometimes it does this by making you yawn. This is also a sign that you are sleepy!

Chapter 5 Nervous system

Your body can sense the world around it in many different ways. It does this through the five senses:

- sight
- hearing
- smell
- taste
- touch.

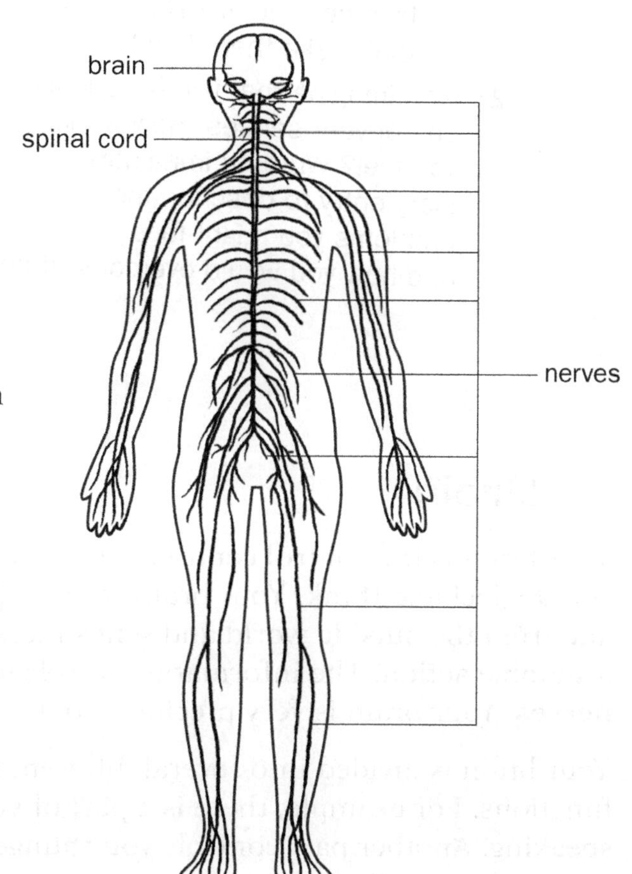

Your body senses information from the world around you through your eyes, ears, nose, tongue and skin and sends signals to your brain. Your brain translates the signals and sends out new signals to other parts of your body. This is called your nervous system.

Your nervous system is made up of your brain, spinal cord, sense organs and a network of nerves that travel all around your entire body.

Activity 5.1 SENSES

How do people survive when they have problems with senses?

1 Interview a person from your community who cannot see or hear.
   - What is their life like?
   - What must they do differently?
   - How have they overcome their disability?
   - How are they treated by other people? How does that make them feel?

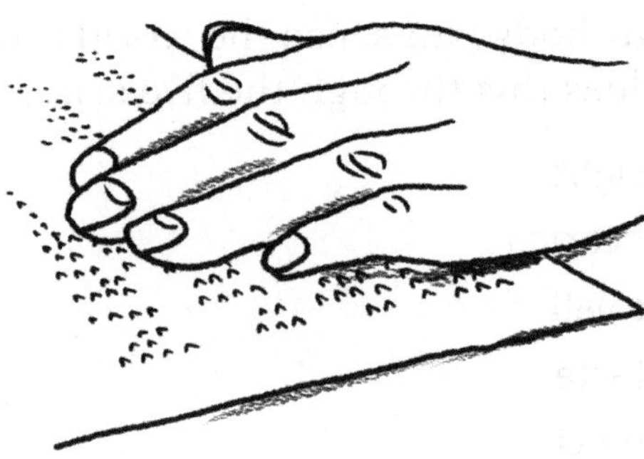

2 Now imagine that you have lost one of your senses. How would you feel? Write an imaginary diary entry to describe the problems you might face and how you would overcome them.

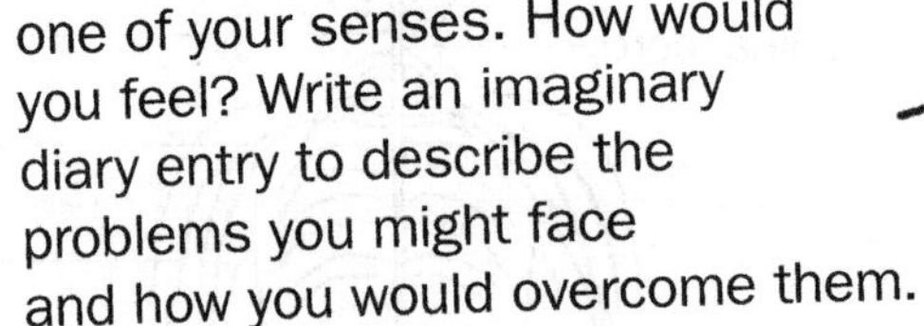

Brain

Your brain is the control centre of your body. It controls everything that you do, feel and think. Your brain receives information from your body and from the outside world and sends messages to your body to perform a suitable action. The information travels to and from your brain along **nerves**. Your brain is very precious, so it is protected by your skull.

Your brain is divided into several different parts which perform different functions. For example, there is a part of your brain that is responsible for speaking. Another part controls your hunger and thirst. The brain is an extremely complicated and fascinating organ and even now doctors and scientists are still learning how it works and thinks.

Human beings have probably the most advanced brain of all the animals. We use complicated tools, tell stories, make music, have a complex language, cooperate and have rich imaginations. Our large brains are one of the things that makes us unique. Even when we are asleep our brains are working – these are dreams.

Did you know?

Your brain is divided into two halves. The left side of your brain controls the right side of your body, and the right side of your brain controls the left side of your body. If you are right-handed, the left side of your brain will control your movements.

Each half of your brain has different skills and abilities. The left side deals with more creative and artistic skills such as drawing and music. It is also where your emotions are processed. The right side of your brain is more logical and deals with decision making, analysis and problem solving.

Activity 5.2 RIGHT- AND LEFT-HANDEDNESS

1 How many right- and left-handed people can you find in your community? Count them and graph the results.
2 How many more right-handed people did you find?
3 What are the advantages and disadvantages of being a left-handed person?

Trouble in the brain: Epileptic fits

Some people have epileptic fits. These are rare and are caused by a problem in the brain. They can cause sudden uncontrolled movement of the arms and legs and often lead to unconsciousness after the fit is finished. It can be scary for both the person having the fit and the people around them.

If you see someone having a fit:

- Stay calm and talk to them.
- Make sure they are safe.
- Put something soft under their head.
- Do not try to stop them moving and do not put your fingers or anything else into their mouth.
- Help them get to a health worker afterwards.

Spinal cord

Your spinal cord is a thick bundle of nerves that connects the brain to the rest of your body. It runs through the tunnel of bones created by your vertebrae (bones in your back). These bones protect the spinal cord from injury.

The nerve of it ...

Nerves are made up of bundles of long nerve cells. An adult human has over 75 000 metres of nerves in their body! What is even more amazing is that some of these nerves are also the longest cells in your body. Your longest nerve runs from the bottom of your spine to the end of your big toe. How long is that?

Nerve cells communicate with other cells using chemicals and the nerve signals travel through the nerve cells as electrical pulses. The fastest nerve signals travel faster than 100 metres per second!

Your body has many different kinds of nerve cells. These include sensory nerves to carry signals such as light and sound from your body to your brain. Motor nerves are another type of nerve cell; these carry signals from your brain to your body to activate your muscles.

Reactions

If you put your hand too close to a fire, nerve cells in your skin are activated. They send a message into your spinal cord and into your brain. You feel pain!

At the same time, an urgent message is passed through the motor nerves that control the muscles in your arm. This message activates the muscle cells, making the muscles contract. The contraction of the muscle will pull your hand away from the fire and protect you from getting burnt. All of this happens in an instant!

This is a simple example of how your nervous system works. It works in a similar way for all kinds of actions and senses.

HOW FAST CAN YOU REACT?

Design a test to measure how fast your nervous system reacts.

For example, drop a ruler between your fingers and measure how quickly you catch it.

Trouble in the brain: Strokes

Older people sometimes have strokes. These are caused by blocked blood vessels in the brain. Strokes can happen in two ways:

- Blood vessels become blocked, stopping the flow of oxygen to the brain.
- Blood vessels burst because of a blockage or weakened vessels.

The blockage is fat caused by an unhealthy diet.

You can tell if someone might be having a stroke because they might get a paralysed arm or face, they might struggle to speak and they might become very confused.

People who survive a stroke can recover well with help.

Trouble in the brain: Fainting

Most of us will faint at least once in our lives.

Fainting is caused by a drop in blood pressure in the brain. This might be because you are dehydrated, you are too hot or you have been standing for a long time. A sudden shock or injury can also cause fainting.

If you see someone faint:

- Stay calm.
- Check they are breathing.
- Check they are safe.
- Lie them down.
- Raise their legs.
- Keep them cool.

 Healthy vision

Many people will have problems with their vision or hearing during their life.

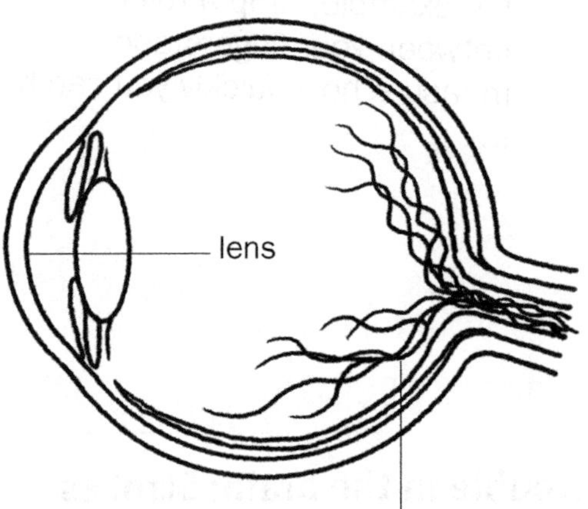

In the middle of the eye, behind your pupil, there is a rubbery lens. This focuses the light coming into your eye. For some people, as they get older the lens changes and it becomes more difficult to see objects that are close to them. This is called being long-sighted.

Some people have the opposite problem. They cannot see objects that are a long way away. They are short-sighted. This is more common in young people and can make it difficult for them to read the blackboard at school. It is important you tell your teacher, parents and health worker if you cannot read the blackboard from the back of the classroom.

Luckily both these problems can be easily fixed by wearing glasses or contact lenses (tiny plastic lenses placed on the eyeball). The lens in the glasses corrects the problem. People should try to have an eye test once every two years because a person's vision changes as they age.

Activity 5.4 FAIR TESTING

How do health workers measure sight?

1 Visit your local health centre and ask the health workers how they measure the vision of patients.
2 Now design a fair test which will measure how well people see.
   - How will you use improvised materials to make the test?
   - How will you make the test fair?
   - How will you measure people's vision?
3 Carry out your fair test on several students, several teachers and several older people. What do you notice?
4 Now ask a short-sighted person to do the same test with and without their glasses. What do you notice? Try on their glasses. What do you observe?
5 Finally, ask a long-sighted person to try the test with and without their glasses. What do you notice? Try on their glasses. What do you observe?
6 Write up your report.

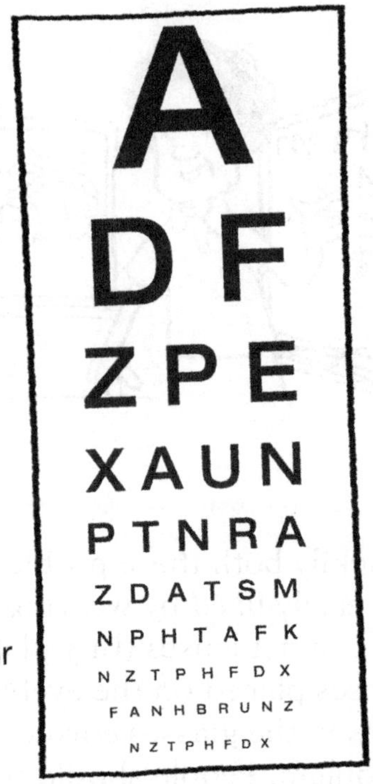

Now design a fair test for hearing.

A
D F
Z P E
X A U N
P T N R A
Z D A T S M
N P H T A F K
N Z T P H F D X
F A N H B R U N Z
N Z T P H F D X

Healthy hearing

Ear infections are common in children in the Pacific. You have probably had one at some point in your life. They are caused by swimming in dirty water. Bacteria get inside the ear and cause an infection. If the infection is not treated by antibiotics from the health worker, the eardrum can become damaged.

Most people do not realise their hearing is getting worse and this can be a serious problem for students. If you cannot hear well, you must tell your teacher, parents or a health worker. Move to the front of the class so you can hear clearly.

Modern technology also causes problems. Listening to loud music through headphones can quickly damage the fine organs inside the ear. A good check to see whether the music is too loud is to ask people standing next to you whether they can hear the music too. If they can, it is too loud and is probably damaging your hearing.

Older people may also have hearing problems.

Chapter 6 Reproductive system

The reproductive system is designed for sexual intercourse and producing new human beings. Without the reproductive systems of your mother and father you would not be alive to read this book.

The organs of the reproductive system are the same as any other body parts and there is no need to feel shy talking about them. It is important to know the names of the different organs and how they work so that you can talk easily with health workers if you need to.

 How new humans are made

Human beings are made when one cell from a man and one cell from a woman join together during sexual intercourse.

After **puberty** a man begins making sperm. Each sperm contains half the DNA needed to make a baby. An adult man produces millions of sperm cells every day in his testes.

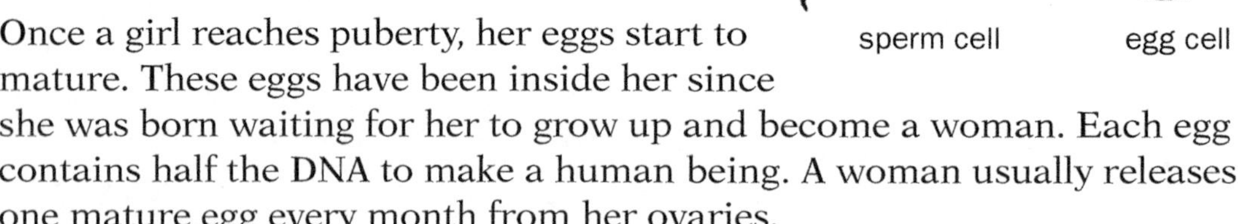

sperm cell egg cell

Once a girl reaches puberty, her eggs start to mature. These eggs have been inside her since she was born waiting for her to grow up and become a woman. Each egg contains half the DNA to make a human being. A woman usually releases one mature egg every month from her ovaries.

During sexual intercourse between a man and woman the man's penis fits inside the woman's vagina. When he is sexually excited the man's penis pumps millions of sperm cells into the woman's body. These travel through the vagina and if they meet an egg, **fertilisation** can happen. The DNA in the sperm cell and the DNA in the egg cell join together and a new human is made.

The new cell starts to divide and over nine months a baby develops and grows inside the mother's uterus (womb). After nine months, the baby is born. Every one of us was made in this way!

Activity 6.1

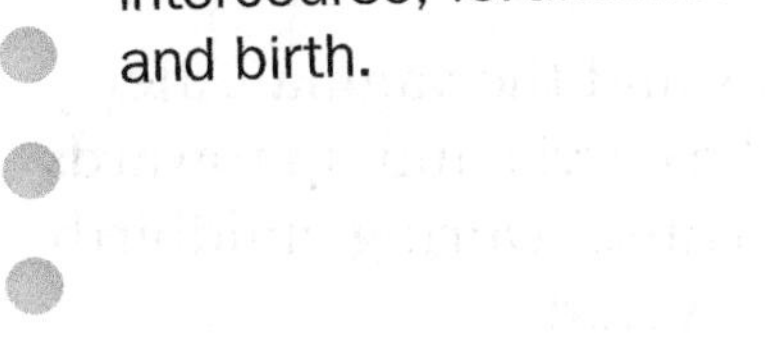

FLOW CHART

Draw a poster showing the life cycle of a human being including sexual intercourse, fertilisation and birth.

The female reproductive system

The female reproductive system is mostly hidden inside a woman's body to protect the precious cargo of eggs and, later, the growing baby.

Inside the female body are several special organs which are needed for making new human beings.

Ovaries

Women have two ovaries. They store thousands of eggs. One egg is usually released every month. This is called ovulation.

Fallopian tubes or oviducts

These are narrow tubes to the uterus. When a mature egg is released from the ovaries it travels down the Fallopian tubes. These are lined with fine little hairs that push the egg towards the uterus. This is where fertilisation usually takes place.

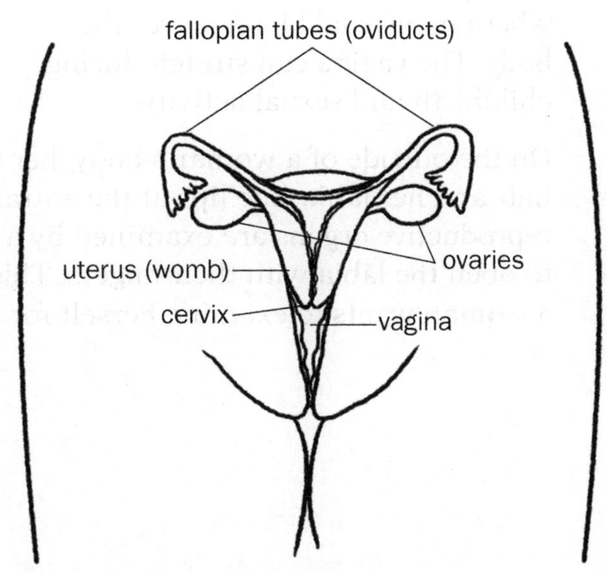

Uterus (womb)

The uterus is a strong muscular bag. It is the place where a baby grows and develops during pregnancy. The growing baby is attached to the uterus by the placenta. During pregnancy the uterus can stretch to almost twice the size of a volleyball! If an egg is not fertilised by a sperm cell, the lining of the uterus is gently shed. This natural monthly bleeding is known as a woman's period.

Cervix

This is the narrow opening between the uterus and the vagina. After sexual intercourse the sperm swim through the cervix and up towards the Fallopian tubes searching for an egg to fertilise. During childbirth the cervix stretches to allow the baby to be delivered.

Vagina

The vagina is a narrow muscular passage leading from the outside of the woman's body to the cervix. The vagina is moist and the amount of fluid can vary throughout a woman's menstrual cycle or when she is sexually excited. During sex, a man puts his penis inside the vagina. Babies are delivered through the vagina and this is also where menstrual blood leaves the body. The vagina can stretch during childbirth and sexual activity.

Healthy bodies

The vagina of a young woman or girl is more delicate than the vagina of an older woman and can be easily damaged by sexual intercourse.

This makes young women more vulnerable to being infected with STIs and HIV.

On the outside of a woman's body, her sexual organs are hidden by pubic hair and her labia (the lips at the entrance to the vagina). When women's reproductive organs are examined by a health worker they will need to open the labia with their fingers. This also happens in birth and if a woman wants to examine herself for infections.

Vulva

The vulva is the name for the external genitals of a woman. It includes the labia and clitoris.

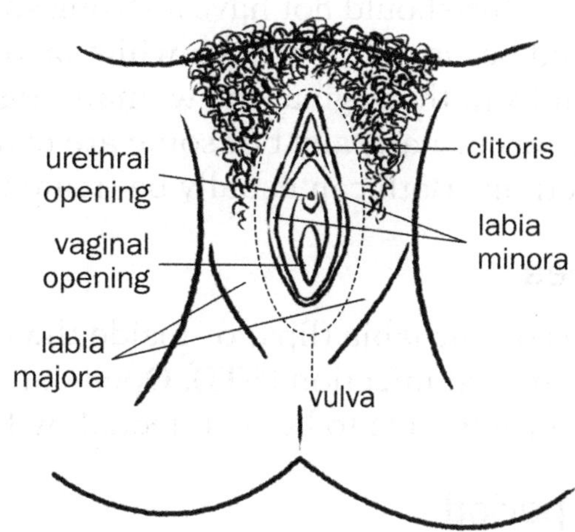

Labia

The labia are fleshy lips that protect the vulva. They vary in size and shape from woman to woman. Pubic hair grows around them after a girl reaches puberty.

Clitoris

The clitoris is a small sensitive organ at the top of the vulva. Gently touching the clitoris increases sexual feelings and pleasure which can lead to an orgasm. The only function of the clitoris is for sexual pleasure.

Urethral opening

This is a tiny hole where the urine comes out.

Vaginal opening

This is the entrance to the vagina.

Keeping the female reproductive system healthy

Keeping the reproductive system healthy is important for a woman's health and if she wants to have a family when she is older. Because a woman's reproductive organs are protected inside her body, it is important for her to recognise any problems early.

Heavy bleeding and pain

If there is more bleeding than usual during a period or there is pain inside her lower abdomen, a woman should always see a health worker. It could be an infection or a problem with her uterus.

Smells and discharges

The vagina should not have a strong smell and the natural fluid it produces should be clear. If there is a white or yellow fluid and a strong, unpleasant smell from the vagina, the woman might have an infection. Sometimes these occur naturally but some are caused by sexually transmitted infections. Both can usually be treated quickly by a health worker.

Sores

Sores on the labia (lips) or inside the vagina might be a sign of a sexually transmitted infection (STI). Open cuts and sores also mean a woman is more vulnerable to being infected by HIV.

No period

Women sometimes have no period (bleeding) during a month. There can be many causes for this, from stress to diet to being pregnant and breastfeeding. There is very rarely a serious problem. Older women (between 40 and 60 years) eventually stop producing eggs and having monthly periods. This is called menopause.

To keep your reproductive system healthy you should do the following things:

- Learn how your reproductive system works.
- Wash with clean water and do not use soaps in the vagina (it is self-cleaning).
- See a health worker if you have any worries or if you have pain, heavy bleeding or unpleasant smells. It could be an STI which might lead to infertility and other complications.
- Wipe your bottom from front to back.
- Wear clean cotton underwear every day.
- Use clean tampons or towels during your period.
- Never leave a tampon inside your body for more than four to eight hours.
- Always use a male or female condom if you have sex.
- Plan your pregnancies and use family planning methods to space children.

Activity 6.2 KEEPING YOUR REPRODUCTIVE SYSTEM HEALTHY

Write an advice leaflet for young women on how to keep their reproductive system healthy.

The male reproductive system

Most of the male reproductive organs are outside of the body, which makes it easy for young men and boys to examine them and keep them healthy.

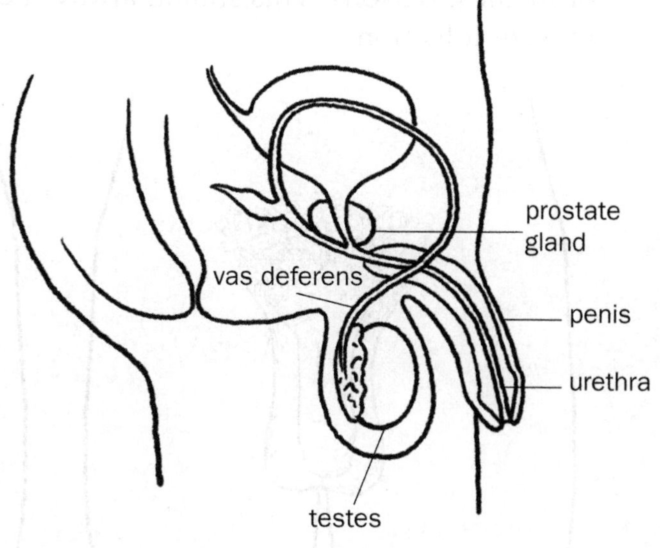

Penis

Penises vary in shape and size. The penis is for passing urine, ejaculation, and for feeling pleasure. It is made of special tissue that can fill up with blood to make the penis larger. This is called an erection. During sexual intercourse, a man puts his penis inside the woman's vagina.

Testes

The testes make sperm cells and hang outside the man's body in a special bag called the scrotum. This is to keep the sperm-making cells cool.

Vas deferens

The vas deferens is the place where the sperm cells are stored before they are released during sexual intercourse.

Prostate gland

The prostate gland produces fluid that the sperm cells swim in. This fluid is called semen. The prostate is a sensitive organ that acts like a pump to push semen out of the penis when a man has an orgasm.

Urethra

The urethra is the tube that runs through the penis which carries urine or semen.

Circumcision

Penises can be either circumcised or uncircumcised. At birth, all men are born with a foreskin on the head of the penis. The foreskin protects the tip of the penis. Some boys and men have the foreskin cut away for cultural or medical reasons. This should always be done by a health worker to prevent infection.

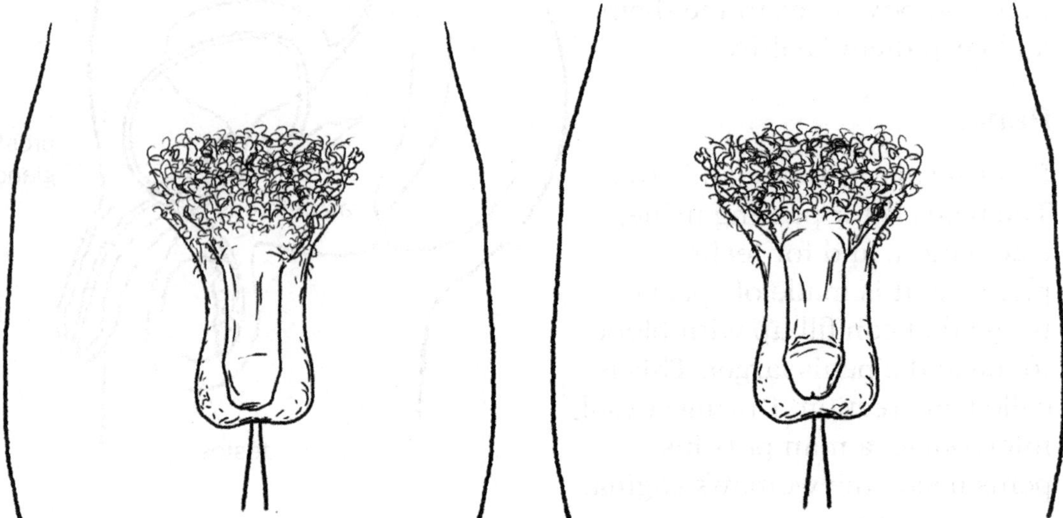

An uncircumcised penis: The foreskin can be gently pulled back for washing and during sex.

A circumcised penis: The foreskin has been cut away.

Keeping the male reproductive system healthy

It is important for boys and men to keep their reproductive system healthy if they want to have children in later life.

Unpleasant smells and discharge

These are a sign of an infection under the foreskin or inside the urethra. Sometimes these infections occur naturally but sometimes they can be caused by STIs. They are usually easily treated with medicines.

Pain when urinating or ejaculating

This is another sign of an infection.

Sores or boils on the penis or scrotum

These can be caused by STIs and increase the risk of getting other infections, such as HIV.

To keep your reproductive system healthy you should do the following things:

- Learn how your reproductive system works.
- Wash under the foreskin every day if you have not been circumcised.
- See a health worker if you have any worries or if you have pain, discharge, bleeding or unpleasant smells. It could be an STI which might lead to infertility and other complications.
- Do not make skin cuts or insert objects under the skin of the penis. Cuts and sores increase the risk of STIs and HIV.
- Wear clean underwear every day.
- Always use a male or female condom if you have sex.

Activity 6.3 KEEPING YOUR REPRODUCTIVE SYSTEM HEALTHY

Write an advice leaflet for young men on how to keep their reproductive system healthy.

Activity 6.4 RESEARCH AND DISCUSSION

Research HIV and STIs in Papua New Guinea. Use textbooks and other reliable information sources to help you (for example, the *Health for the Pacific* textbook on HIV/AIDS and STIs). Answer these questions:

1. Why is it important that young men and women learn about HIV and STIs?
2. Why is it important that students learn about their reproductive systems and how they work?

List at least five other questions you would like to ask about the human reproductive system. Work with your peers and teacher to find the correct information.

Chapter 7 Digestive system

Every day we need to eat and drink to remain healthy and grow. Human beings have a complicated system for eating and drinking called the digestive system. Almost all of your digestive system is hidden away inside your body, but you can sometimes hear, feel and smell it working.

Activity 7.1 FOOD DIARY

How much food do you eat every day? How much do you drink?

1 Keep a food diary for one week. If you can, try to measure or estimate the weight of the food you eat and drink.

2 At the end of the week, calculate how much food you have consumed in kilograms. How much do you think has been used by your body?

3 Compare your diary to a friend's. What is the same and what is different? Why?

Your body performs many functions every day, such as breathing, thinking and moving. Your body also repairs itself, keeps your temperature constant and makes new cells to replace damaged ones. To do all of this your body needs food and drink to provide energy. Your body breaks down food and drink into vitamins and minerals that it can use to strengthen and keep your body functioning in a healthy manner.

Super-sized me

Your digestive system is very long indeed! In adults it can stretch over nine metres. Measure this out in your classroom.

The reason it is so long is so that your body can extract everything it needs from your food. In a healthy person, it will take between 24 and 72 hours for your digestive system to process a meal that you eat.

The digestive system

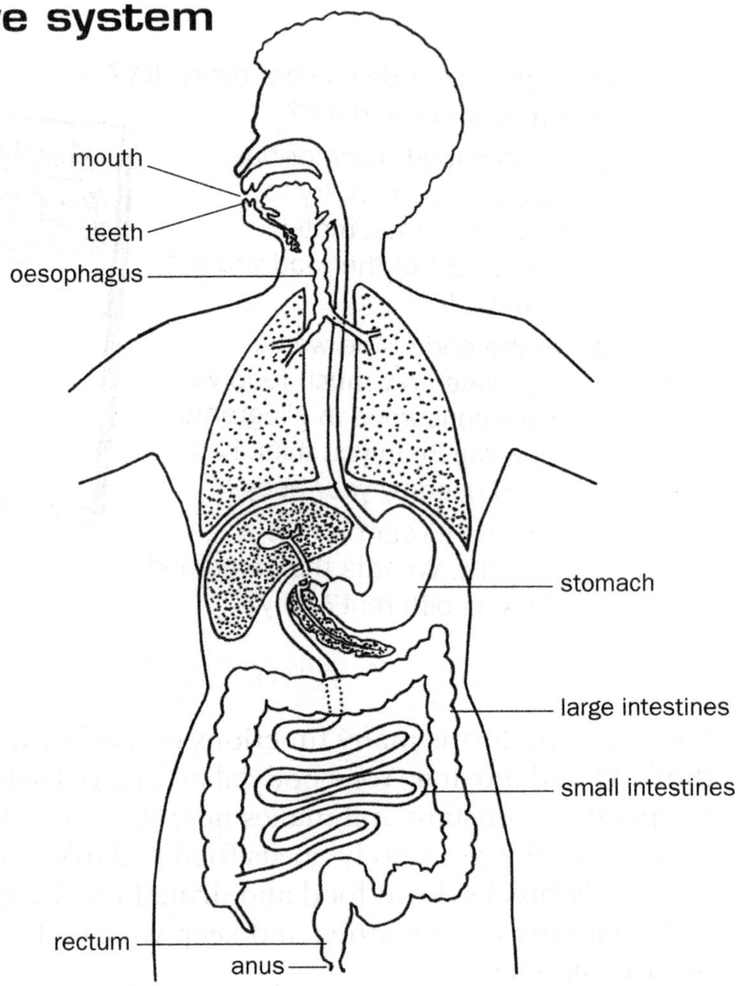

mouth
teeth
oesophagus
stomach
large intestines
small intestines
rectum
anus

Mouth

Your teeth tear and crush the food that you eat. Throughout your childhood you will lose your first set of teeth. These will be replaced by your 32 adult teeth. If you lose your adult teeth they will not grow back, so it is very important to take good care of your teeth. Drinking sugary drinks and chewing betel nut damages teeth. Brush your teeth at least twice a day with fluoride toothpaste.

Saliva contains special chemicals which dissolve the food and help it travel through the digestive system. Your body makes one and a half litres of saliva a day!

After chewing, you swallow the food ...

Oesophagus

The oesophagus is the tube that connects your mouth to your stomach. Food is swallowed and forced down the tube by muscles in the oesophagus.

A special flap called the epiglottis stops the food from entering the lungs. Another valve stops the stomach acid from escaping up the oesophagus.

Stomach

Your stomach is a pouch which stores and helps to break down food.
The stomach walls are made of thick elastic muscles. When your stomach is full it can hold almost four litres of liquid.

The stomach contains digestive acids and chemicals which break down the food that you eat. For most people, the food that you eat will spend about three hours in your stomach.

The stomach lining absorbs vitamins and other small molecules such as alcohol. These pass through the stomach lining and into the blood.

When food leaves the stomach it has been broken down into a partly liquid form.

Small intestine

Food travels from your stomach into your small intestine. The small intestine is a long muscular tube that is coiled up inside of your body and is about four metres long. It is called the 'small' intestine because it is only about four centimetres wide. Muscles in the small intestine contract to squeeze food along.

This is where most of the digestion and absorption takes place. The small intestine produces different kinds of liquids to further break up food and helps absorb the food into the blood.

The lining of your small intestine is covered with small hair-like structures which improve absorption of different molecules.

Detour!

Some of the food that you eat requires special natural chemicals called enzymes to break it down. These chemicals can be found in different organs that are linked to the small intestine.

Food molecules travel through the blood from the small intestine to the liver and pancreas where they can be further broken down.

Your liver produces a liquid called bile. It breaks down fats as food passes through your small intestine. Bile is stored in your gall bladder until it is needed. When bile breaks down fats, some of the materials are absorbed into your bloodstream for your body to use. The waste materials are returned to the small intestine.

Your pancreas is another small organ that produces enzymes that break down fats, proteins, carbohydrates and chyme. Your pancreas secretes the enzymes into the small intestine to help with digestion.

However, a lot of the hard work to break down your food is actually done by millions of friendly bacteria who live in your intestines. Your body gives them a safe and warm place to live. You can sometimes smell the gases these helpful bacteria make as they are working, when you break wind (fart)!

Large intestine

After the small intestine, what is left of your food enters the large intestine. It is only about one and a half metres long but much wider than the small intestine.

The large intestine absorbs water from food. Other waste products from the body enter the large intestine and these are mixed with any leftover or unwanted food to become concentrated faeces. In fact, the brown colour of your faeces is caused by the old dead red blood cells your body removes.

Faeces is stored in the rectum and then regularly passed out from the body through the anus.

Activity 7.2 STORY WRITING: A DAY IN THE LIFE OF A BANANA!

Write the story of a banana from the moment it is eaten to when it is excreted as faeces.

Start your story with these sentences:

> *I was hanging there on the tree in the sunshine when a human being walked over. She picked me and ripped off my skin. The next thing that happened was ...*

Use your knowledge of the digestive system to explain what will happen to the banana as it is digested.

Leftovers!

Your **appendix** is a small pouch connected to your large intestine. It is a dead end and does not seem to have any function at all in humans.

Doctors and scientists believe that the appendix used to help humans digest plant food. Over thousands of years the foods that we eat have changed, and are easier to digest, so the appendix is no longer needed. It is a leftover!

Unfortunately your appendix is still a part of you and can become infected. If your appendix gets very infected you might get appendicitis. Your appendix will swell and there will be pain in your lower stomach on the right side. If this happens you should see a health worker immediately. If your appendix bursts inside you, it can be very dangerous.

 Healthy diets

Different foods contain different vitamins, minerals and nutrients, which are all things that you need to keep healthy and to grow. It is important to eat a balanced mix of foods so that your body can absorb all of the different nutrients.

Some examples of these nutrients are shown in the table below.

Nutrient	Use of nutrient in your body	Foods that the nutrient is found in
Carbohydrate	Energy	Bread, potatoes, kaukau, rice, sugar
Fat	Energy and warmth	Butter, milk, oil
Fibre	Healthy digestion	Fruit, vegetables, whole wheat bread
Protein	Growth and repair	Fish, milk, nuts, lean meat
Vitamins and minerals	Control chemical processes	Fruit, vegetables, fish, milk, eggs

Balanced diet

It is recommended that healthy adults should eat:

- 2–3 portions of meat and fish
- 2–3 portions of dairy
- 4–5 portions of fruits and vegetables
- 6 portions of potatoes, bread and rice.

Energy from food is measured in kilocalories (kCals) or kilojoules (kJ). Healthy adults need approximately 8700 kJ each day to stay healthy and to have enough energy.

Activity 7.3 KNOW WHAT YOU'RE EATING!

Today, many food labels have a table that lists the amount of nutrients and number of kilojoules that are contained in that food item. Most labels will list proteins, carbohydrates, fats, sugars, fibre, sodium (salt) and different vitamins and minerals.

1 Collect food labels from different foods.
2 Compare the amount and different kinds of nutrients that are found in different foods.
   a What kinds of foods contain the most sodium (salt)?
   b What kinds of foods contain the most sugar?
   c What kinds of foods contain the most protein?
   d What kinds of foods contain the most fat?
   e What kinds of foods contain the most kilojoules? Are you surprised?

Unhealthy digestive systems

Not eating enough food or the right kinds of foods can lead to many problems. These can range from simple to very serious and life-threatening.

Low energy and tiredness

This can be caused by not eating enough food. If you do not eat enough you will feel tired and unable to concentrate. This will make it difficult for you to play sports or pay attention and learn in school. Coming to school after eating a poor breakfast does not help your learning.

Vitamin and mineral deficiencies

Vitamin deficiencies can cause problems in the body. For example, not having enough vitamin D means your bones do not grow properly.

Mineral deficiencies can cause problems in the body. For example, not having enough iron leads to anaemia.

Obesity

Obesity is caused by eating too much fatty food, which leads to you becoming overweight. Over the years, too much fat in your diet blocks your arteries and can cause heart attacks and strokes. People who are overweight have serious extra risks.

Diabetes

Diabetes is caused by too much sugar in your diet. The extra sugar burns out the cells that control the amount of sugar in your blood. Diabetes is a serious problem in the Pacific. Soft drinks and cordial contain a lot of sugar. Many people have too much sugar in their tea.

Constipation

Constipation is caused by not eating enough fibre (greens, vegetables and fruits) and not drinking enough water. This is dry faeces stuck in the rectum and is very uncomfortable.

Worms

Worms are transmitted by animals and faeces. Many children in the Pacific have intestinal worms. The risk from worms can be reduced by drinking clean water, washing hands with soap and having safe toilets. Children with worms do not do as well in school because the worms use up their food and nutrients. One sign a child is infected with worms is that they have a very itchy bottom.

Diarrhoea

Diarrhoea is a result of infections in the digestive system caused by poor hygiene. Diarrhoea is a common cause of death in babies and infants. It can be prevented by cooking food well, keeping flies off food, washing hands with soap after using the toilet and before cooking, and having safe water and good toilets. If someone has diarrhoea they must drink rehydration solution.

Cholera

Cholera is a very dangerous infection caused by dirty water and poor hygiene. It leads to extreme diarrhoea which can kill people quickly unless treated with rehydration solution. If you think you have cholera you must see a health worker.

Activity 7.4 DO YOU HAVE A HEALTHY DIET?

Keep a food diary for one week. Record what you eat and drink at each meal for seven days.

	Sunday	Monday	Tuesday	Wednesday	Thursday	Friday	Saturday
Breakfast							
Lunch							
Dinner							
Snacks							
Other							

1. Did you eat healthy foods?
2. Did you eat unhealthy foods?
3. Do you eat a balanced diet?
4. Did you eat too much of one kind of food?
5. Compare your food diary with your friends' and then write a healthy diet action plan.

Tooth decay

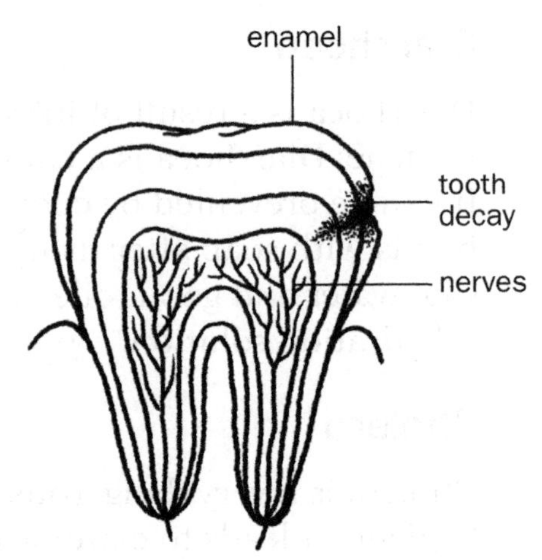

Tooth decay is a common problem in PNG. Teeth become damaged by drinking too many sugary drinks and chewing betel nut with lime. Betel nut and cigarette smoke also cause stained teeth. Bacteria live on the sugar and eat into the teeth. Tooth decay is very painful and can lead to losing teeth and getting gum disease (where the gums are sore and bleed).

Dentists can treat the problem by drilling out the decayed part of a tooth and filling the hole. You can protect your teeth by avoiding sugary drinks, not chewing betel nut, using dental floss and brushing your teeth with fluoride toothpaste after meals and before going to bed.

Chapter 8 Excretory system

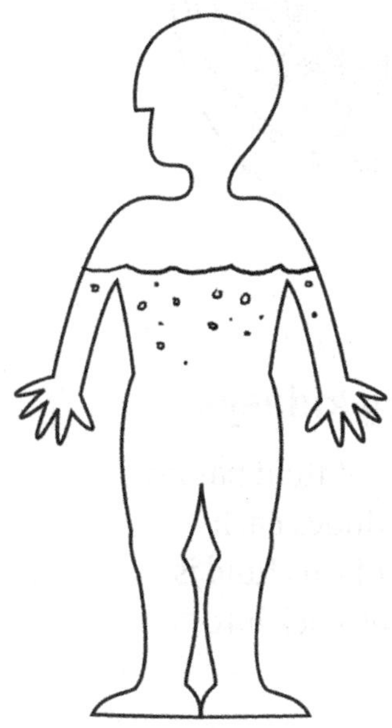

Your body is made of many different materials. The most common of these materials is water. Water is found in all of your cells, tissues, organs and bones, and is used in many of the different processes in your body. In fact, two-thirds of your body is made up of water.

Your body's excretory system controls the level of water in your body to keep you healthy. If you take in more water than you need, your body can get rid of the extra liquid.

Your excretory system also gets rid of waste materials from your body. Every day, thousands of chemical processes happen inside your body. These processes produce waste materials that your body must remove.

The major organs that make up your excretory system are the kidneys, ureters, bladder and urethra.

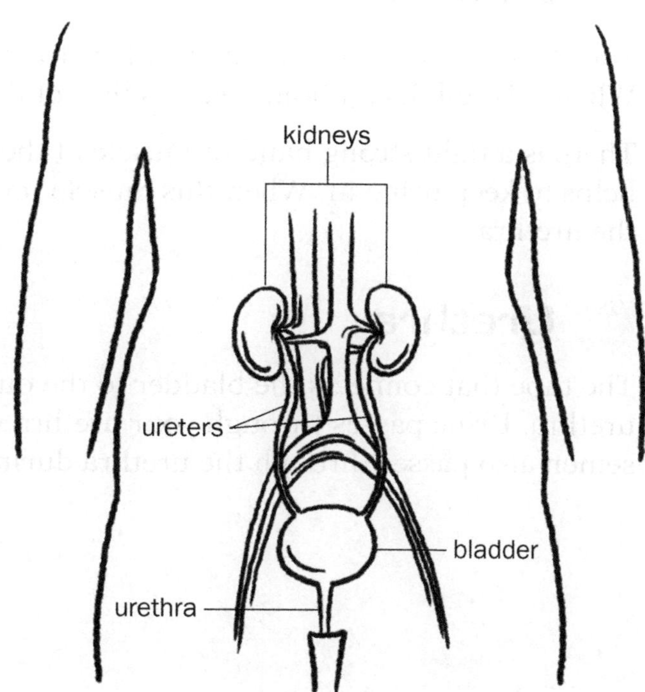

Kidneys

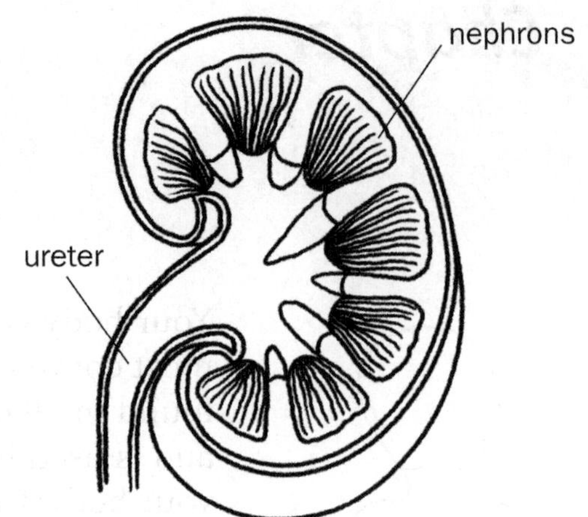

Your two kidneys are located in your lower back on both sides of your spinal column. They are protected by the lower ribs in your rib cage.

Inside each of your kidneys there are more than one million nephrons. Nephrons are small tubes that act as a filter to remove wastes, salt, excess minerals and water from your blood. These wastes are called urine.

Ureters

Urine is carried from the kidneys to the bladder through muscular tubes called ureters. In adults ureters are between 25 and 30 centimetres long. (That's about the length of a ruler!)

Hard-working kidneys

About 150 litres of fluid passes through your kidneys each day. Almost all of this fluid is cleaned and goes back into your blood.

Bladder

Your bladder is a muscular bag which stretches as it fills up with urine. When it is full, it can hold nearly a litre of liquid.

There is a tight strong band of muscle at the neck of the bladder that helps to keep urine in. When this muscle relaxes, urine flows into the urethra.

Urethra

The tube that connects the bladder to the outside of the body is the urethra. Urine passes through your urethra when you pee. In men, semen also passes through the urethra during sex.

A rainbow of colours

Your urine (pee) can tell you many things about your health and the food you have been eating. The colour and smell of your pee changes depending on the amount that you drink or what you have to eat and drink.

- Urine that is clear or has no colour is a sign that you are drinking lots of liquids. Your body may have too much liquid.
- Urine that is dark yellow may be a sign of dehydration. You are not drinking enough liquid.
- If there is blood in your urine, this may be a sign of an injury or infection. You should see a health worker straight away.
- Eating beetroot can cause your urine to have a pink colour and eating asparagus can cause your urine to turn green!

The smell of your urine can also be affected by the different foods that you eat. Drinking alcohol or coffee and eating asparagus or onions can all change the way that your urine smells.

More water please!

When a person does not drink enough liquid they risk dehydration. Dehydration is common when people are sick with diarrhoea or vomiting. You can also get dehydrated when playing sport or working in the garden in the hot sun.

Dehydration can happen in people of any age, but it is very serious and dangerous in small children.

Signs of dehydration include:

- thirst
- little or no urine
- very dark yellow urine
- weight loss
- dry mouth.

To prevent or treat dehydration you should:

- drink plenty of clean water and *kulau*
- drink rehydration solution
- lie down in a cool place.

Very dehydrated people and children will need to have a drip inserted at a health centre.

Activity 8.1 REHYDRATION AT HOME

Dehydration can be very dangerous. Luckily, it can also be prevented and treated very easily and quickly. You can get 'ORS' (oral rehydration salts) and other rehydration solutions at the chemist or from your health worker. Sometimes you might be too far from town or the nearest aid post, but you can make your own rehydration solution at home. With your parent or guardian, prepare an emergency rehydration kit for your house.

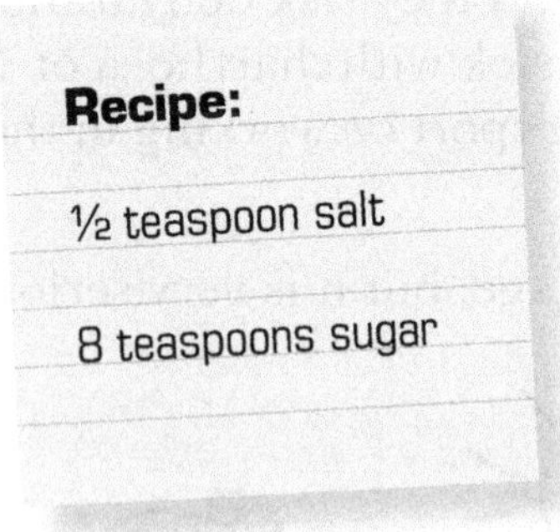

Recipe:

½ teaspoon salt

8 teaspoons sugar

Mix the salt and sugar together and store in a sealed container in a cool and dry place with an empty one-litre container.

When you need to rehydrate, put the salt and sugar mixture into the one-litre container, fill it with clean water, and mix until the mixture dissolves. Drink as much of this as you can until you start to feel better. If you do not feel better after two days, you should see a health worker as soon as possible.

Rocky ride!

If your kidneys do not get enough liquid passing through them, the urine they make might be very concentrated. Solid material in the urine can join together to form kidney stones.

If the stones are bigger than peas they can block the ureters and cause terrible pain. If you think that you or someone you know is suffering from kidney stones you should see a health worker immediately.

Activity 8.2 YOU ARE WHAT YOU DRINK!

- Keep a diary for one week. Record (in millilitres and litres) how much you drink at each meal and throughout the day for seven days.

	Sunday	Monday	Tuesday	Wednesday	Thursday	Friday	Saturday
Breakfast							
Lunch							
Dinner							
Snacks							
Other							

1. How much do you drink each day?
2. What do you drink?
3. When do you drink?
4. Do you drink when you are hot? Do you drink when you are thirsty?
5. What changes can you make to have a healthy and balanced diet?

Compare your answers with your friends'.

Drinks that make you pee!

Some drinks actually make you urinate more. These include:

- cola
- tea
- coffee
- cocoa.

The effect is caused by the caffeine in the drinks.

It is not a good idea to drink these drinks if you are feeling thirsty. Water and *kulau* are much better at quenching thirst.

Sugary drinks

Drinks such as cola, cordial and lemonade contain huge amounts of sugar. Too much sugar in your diet can make you fat and cause diabetes.

How much sugar is in these drinks?

Read the side of the bottle or can. For example, one can of lemonade contains 40 grams of sugar! It is better to drink water, *kulau* or sugar-free drinks instead.

Activity 8.3 SUGAR, SUGAR EVERYWHERE!

Your body is made up mostly of water. Water is found in all of the cells, tissues, organs and fluids that make up your body. How much you drink and what you drink can keep you healthy. The best things to drink are water and *kulau*. However, in the Pacific, many people like to drink cordial and soft drinks. These taste good and are OK to drink in small amounts, but these drinks are filled with sugar and salt. If you drink a lot of these, they can damage your health.

Do this activity with your friends or class. You will need:

- 1 glass of water
- 1 glass of soft drink
- 2 chicken bones or old teeth.

1. Put one clean chicken bone or old tooth into the glass of water and put one clean chicken bone or old tooth into the glass of soft drink (e.g. cola or cordial).
2. Store both glasses in a safe, cool place.
3. Each day, remove the bones or teeth from the liquid and check them. What do you see? Are they changing? How are they changing? What do you think is happening?
4. Do this for ten days. Keep a record of the changes that you see each day. Compare with your classmates.
5. You have seen what soft drink can do to a hard material. What do you think it will do to your teeth?

Healthy water

Many communities in the Pacific do not have clean water supplies or good toilets. Schools might not have enough toilets for their teachers and students and rarely provide soap. Many young people and children get sick because of this. Without good sanitation, students will defecate and urinate in and around the school grounds.

Safe sanitation

Toilets should be located away from fresh water sources and should be safe for girls and young women to visit. Many communities are starting to use ventilated pit latrines, which smell less than pit toilets and are simple to build.

Ventilated pit latrine

Students should *always* use soap or wood ash to scrub their hands after using the toilet. Toilet paper should be available for them to use.

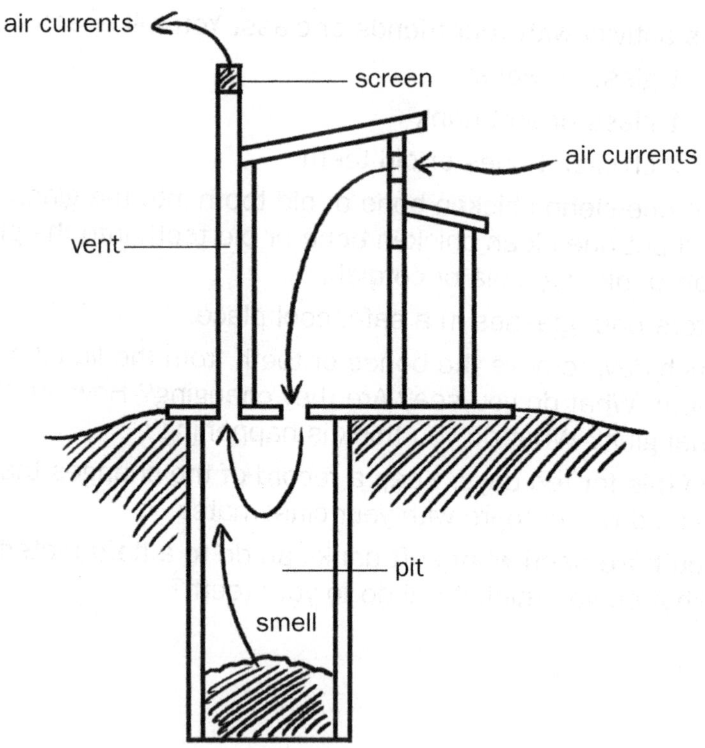

Safe water

Clean water is a basic human need. Water sources should be protected from animals and human faeces. Communities and schools should install water systems or water tanks as a priority.

Activity 8.4 HEALTHY SCHOOL AND COMMUNITY CHECKLIST

1 With a partner, design a healthy school and healthy community checklist for water and sanitation (toilets). For example:

School checklist

☑ Enough clean functioning ventilated pit latrines or toilets for students (i.e. one toilet for every 25 students)

Number of girls per toilet ____ Number of boys per toilet ____

☑ Doors on latrines with secure lock

☑ Latrines cleaned daily

☑ Toilet roll in each latrine

☑ Soap or wood ash and clean running water available for washing hands after going to the toilet and before eating

☑ Rain water tank or piped water supply with cup for students to drink out of

☑ Mosquito screens on all water tanks

☑ No faeces from animals or babies/small children on school grounds

☑ Towels and pads available for menstruating girls

☑ Is the safe drinking water facility adequate according to the number of students and teachers?

Adapted from: PNG Department of Education Health Promoting Schools checklist, 2009

2 Use the checklist to assess your school and your community. Is your school healthy for water and sanitation? Is your family home healthy?

3 Write your recommendations on how to improve your school and your family area. Who will you give the recommendations to? What will you implement yourself?

Chapter 9 Immune system

Every day when you eat, drink, breathe and move about your community, you come into contact with thousands of tiny living organisms commonly known as **germs**. Germs can cause infections, disease and illnesses. Germs can be:

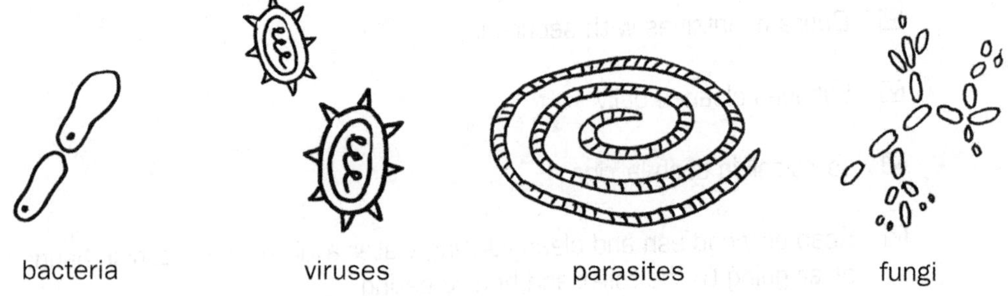

- viruses (like the ones that cause flu, polio or HIV)
- **bacteria** (like the ones that cause food poisoning, pneumonia or tuberculosis)
- parasites (like the ones that cause malaria and dysentery)
- fungi (like the one that causes ear infections).

Germs enter your body through your nose, mouth and ears and through cuts and scrapes. They can be spread by air, water, food and animals or insects, so they are known as **communicable diseases**. Most germs are too small to see without a microscope. When they are inside your body, they multiply and attack your body, causing disease and illness.

Activity 9.1 GERM HUNT

Which germs have you heard of? Brainstorm all the different diseases you know and then sort them into the right group.

Bacteria	Virus	Parasite	Fungi
e.g. tuberculosis	e.g. influenza	e.g. intestinal worm	e.g. grille

To the rescue!

Your body has many different ways to stop germs from getting inside your body. These include:

- **Skin:** your skin tries to keep germs from entering your body and forms the best barrier against germs.
- **Nose:** little hairs and mucus (snot) on the inside of your nose trap the germs and dirt that you breathe in.
- **Ears:** little hairs and wax on the inside of your ears trap germs and dirt and stop them entering the inner ear.
- **Eyelids:** your eyelids and eyelashes keep germs and dirt out of your eyes. Tears help to wash your eyes and keep them clean.
- **Saliva** and **stomach acid:** these kill germs in your food as you eat.

When germs do manage to get inside your body, your immune system helps to prevent infection, disease and illness.

Your immune system is made up of many different kinds of cells. Some of the most important cells in your immune system are the white blood cells. White blood cells make up part of your blood. They travel throughout your body in search of germs and other foreign bodies.

When your immune system meets something from outside the body, such as a virus, it makes small particles called **antibodies**. Antibodies stick to the invaders and help the rest of the immune system to find and destroy the germ to help the body get rid of infections. This allows a person to avoid illness, or to become well if they are already ill. Your body makes a different antibody for each different kind of infection.

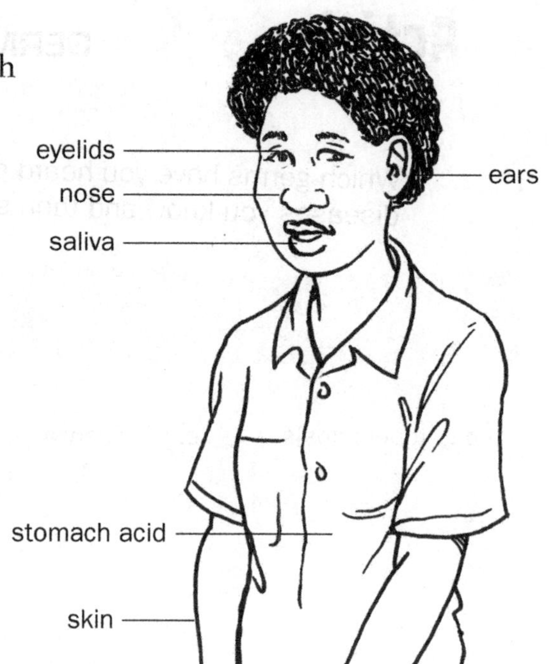

Some white blood cells swallow up germs when they find them. Other white blood cells instruct the body how to fight off the infection.

 Road to health

Your white blood cells are constantly moving throughout your body. Your white blood cells travel around your body in your blood. They also travel around your body through your **lymph** system. Your lymph system is a set of tubes that carries fluid and white blood cells throughout your body.

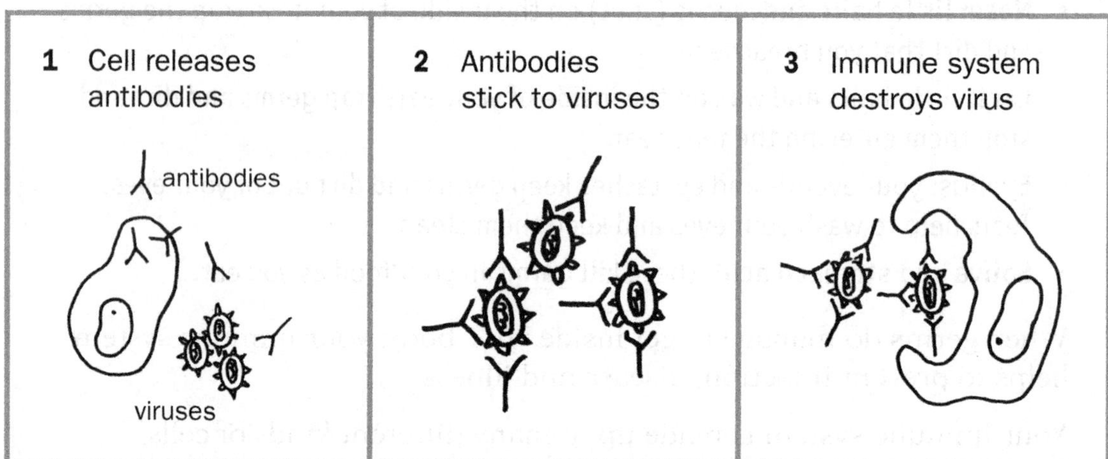

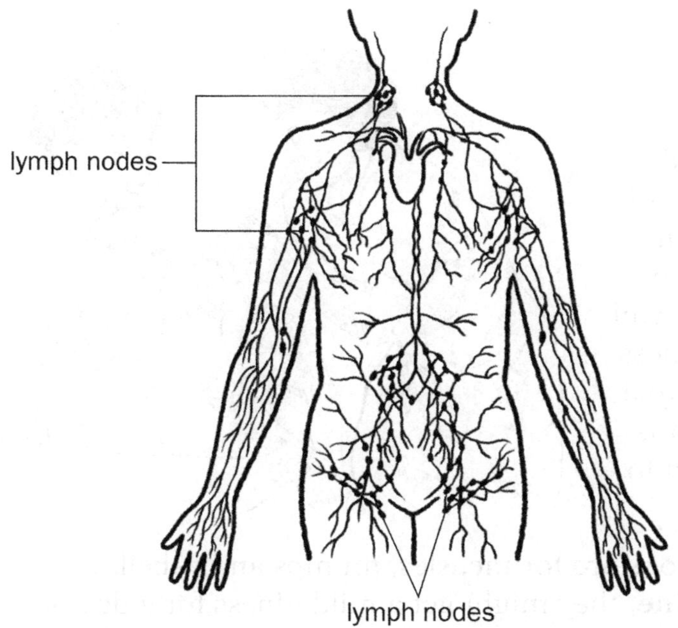

Nodes are small bags that store and produce white blood cells. Lymph nodes are in your neck, armpits and groin. They are usually small. If you become sick, the lymph nodes start to make extra cells and will get bigger so that you can feel them, or even see them. When your lymph nodes swell, it means that your body is fighting an infection.

 Antibody for life!

Once your body has identified a germ and made the correct antibody for that germ, it can make it again and again very quickly. If you happen to get infected with the same germ, your immune system will make the antibody for that germ very quickly and kill the germ before you become sick. Your immune system is constantly learning to fight new infections. Most of the time you will not even realise it is fighting an infection.

Breast is best

Breast milk contains antibodies from the mother's body to help the newborn baby fight off any infections. This is another reason why breastfeeding is best for babies.

Vaccines

Scientists have learned that if they give you a small amount of a weakened germ by injection or drops, your body will make antibodies for that germ and will be able to prevent serious illness if you become infected with that germ in future. This is why it is important for young children to get all of their vaccinations.

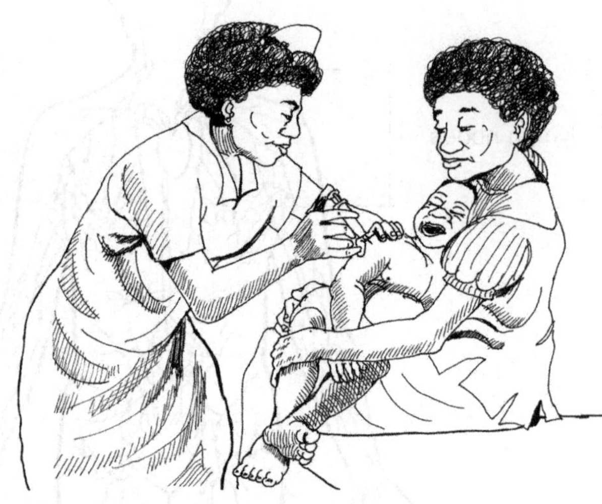

The most common vaccinations are for measles, mumps and rubella. After a child receives a **vaccine**, they might get a mild illness for a day or two. During that time their immune system will develop antibodies for those specific germs.

If the child comes into contact with the germ in the future, they will already have the antibodies in their body and they will be able to prevent serious illness.

Activity 9.2 WHICH VACCINES?

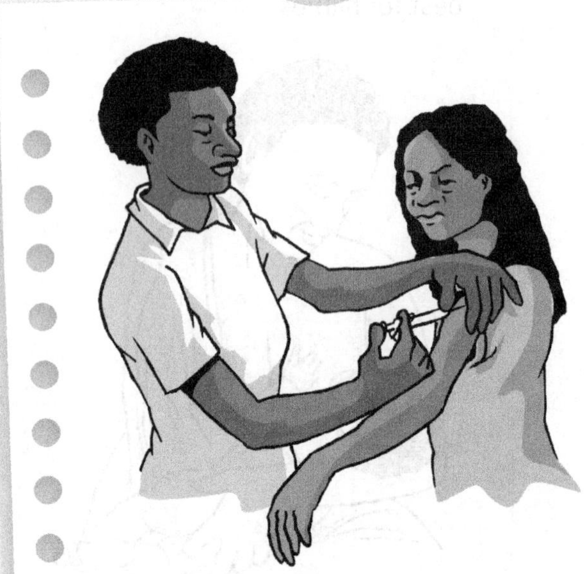

Interview a health worker about which vaccines they give and when. Draw a timeline and mark on it the stages a baby and child should receive a vaccine, and which disease the vaccine helps to prevent.

Keeping your immune system healthy

Your immune system works better if you are healthy and fit. If your diet is not very good and you live in a dirty and unhygienic community you will get more infections.

Activity 9.3 DISEASE CASE STUDIES

Read the case studies below.

- **a** Which germ is most likely responsible for each child's illness?
- **b** Which illnesses would need the help of a health worker?
- **c** How would each illness be treated?
- **d** What would you do to help each child?

Illness #1

The child has a very high fever, is sweating a lot and keeps crying. Her eyeballs are slightly yellowish.

Illness #2

The child has an upset stomach, seems tired all the time and keeps scratching his bottom.

Illness #3

The child has a bad cough and a high fever. She is struggling to breathe. This cough has been going on for a long time.

Illness #4

The child is vomiting violently and has constant diarrhoea. The diarrhoea is very watery.

Illness #5

The child has red swollen eyes. The eyes are making lots of tears and pus.

Illness #6

The child has a sore on his leg which does not seem to be healing. It is very red and inflamed and there is a lot of pus. It is very painful.

Illness #7

The child has quickly developed lots of small red spots all over her body. The child has a fever.

Illness #8

The child has a patch of white dry skin on his arm that itches.

Activity 9.4 DISEASE CASE STUDIES – THE ANSWERS

Remember: symptoms of some illnesses are very similar. If you are not sure or if the symptoms are serious (high fever, constant diarrhoea, etc.) see a health worker *immediately*.

Illness	Probably...	What to do	How to prevent it
1	Malaria	See a health worker immediately. Treat with anti-malarial medicine and fluids. Complete the course of medicine.	Sleep under a mosquito net. Empty or cover containers of standing water where mosquitoes breed. Use insecticide to kill mosquitoes.
2	Intestinal worms	Take anti-worming medicine from the pharmacy.	Wash hands after going to the toilet. De-worm children every six months.
3	Chest infection or (more rarely) tuberculosis	See a health worker immediately. Treat with antibiotics.	Keep house dry. Avoid smoke.
4	Severe food poisoning or (rarely) cholera	See a health worker immediately. Give child oral rehydration solution immediately. Might need a drip.	Wash hands after going to the toilet. Drink clean water and eat well-cooked food.
5	An eye infection called conjunctivitis (also known as red eye)	Take medicine from the pharmacy. Avoid rubbing eyes.	Wash hands.
6	Tropical ulcer	See a health worker immediately. Cover up the sore and change the dressing regularly. Will need antibiotics, but must complete the course.	Wash hands.
7	Measles or chickenpox	See a health worker immediately if the fever becomes worse. Treat the spots if they become infected. Body will eventually fight it off.	Have a measles vaccination.
8	Grille	Treat with anti-fungal cream from the pharmacy. Don't scratch.	Wash body with soap. Check your water source is clean.

Chapter 10 Messaging system

Every day, your body performs thousands of different actions to keep you healthy and active. As you grow and develop from a child to an adult, your body changes. Your body has a system that controls and monitors your body systems and how they are working. It is called the endocrine system.

The endocrine system is made up of many different **glands**. A gland is a small organ that makes and releases different chemicals.

The most important glands in the endocrine system are:

- **Pineal gland:** located in the middle of your brain
- **Hypothalamus:** located in the central part of your brain
- **Pituitary gland:** located at the base of your brain and no bigger than a pea
- **Thyroid gland:** located in the lower front part of your neck
- **Adrenal gland:** located on the top of each kidney
- **Pancreas:** located behind your stomach
- **Ovary** (in women): located on either side of the uterus
- **Testes** (in men): located in the scrotum.

The glands in the endocrine system make chemicals called **hormones**. Each gland makes a different hormone. Hormones act as the body's messengers and transfer information and instructions from one part of the body to another. The hormones are released into the bloodstream by the glands and travel through the bloodstream to other parts of your body.

Your body makes more than thirty kinds of hormones and each hormone has a very specific function. Hormones help to control your mood, how you grow and develop, how your tissues work, and your sexual and reproductive systems.

Some of the hormones act very quickly but some take a long time to have an effect.

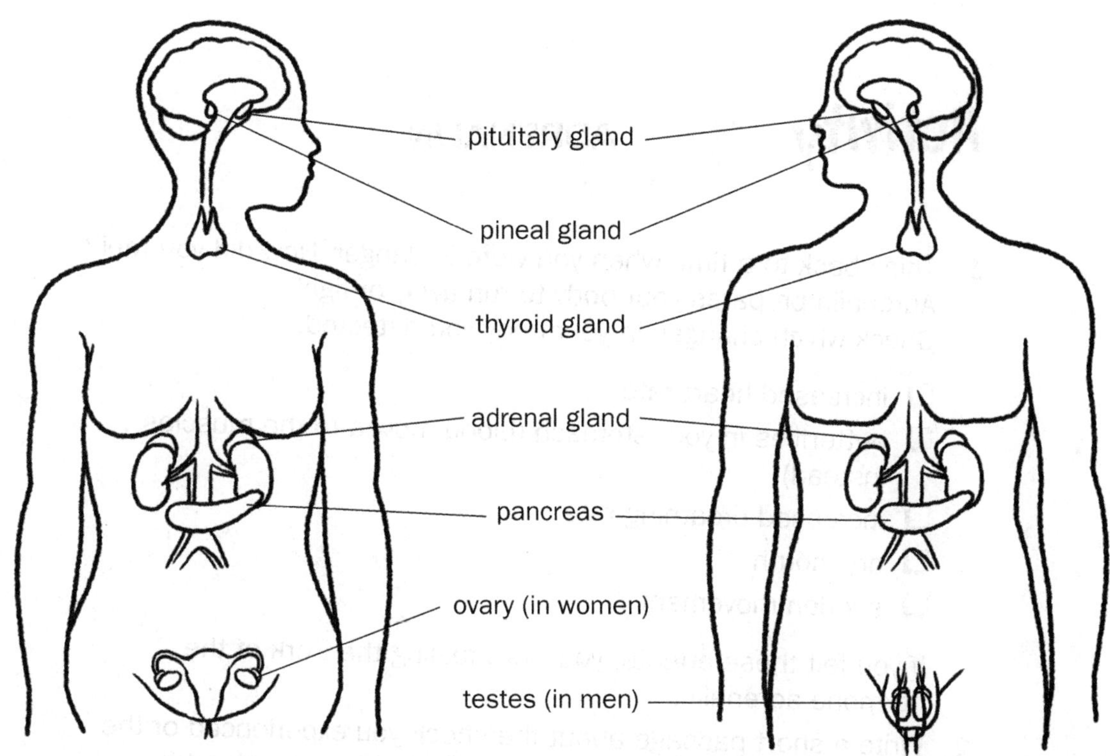

 A few of the most important hormones

Adrenalin

Everyone has experienced the powerful effects of adrenalin. This hormone is produced by your adrenal gland when you have a sudden shock or you are stressed. It is a fast-acting hormone and is sent quickly around your body.

The nickname for adrenalin is the 'fight or flight' hormone. It prepares your body to either run away from the danger or fight to save yourself.

Activity 10.1 ADRENALIN

1 Think back to a time when you were in danger. How did you feel? Adrenalin prepares your body to run away or fight. Check which changes in your body you detected.

- ❑ increased heart rate
- ❑ butterflies in your stomach (blood moved to the muscles instead)
- ❑ increased breathing rate
- ❑ dry mouth
- ❑ sudden movement

If you felt these effects, you were feeling the work of the hormone adrenalin.

2 Write a short passage about the shock you experienced or the danger you were in, how you felt and what you did about it. Did you flee or fight?

3 Have you ever felt the effects of adrenaline in other situations? It might have been when you were playing sports or trying something new. Write a short passage about this situation, how you felt and what you did about it.

Reproductive hormones

As young people grow from children into adults, they experience puberty. During puberty your body changes. These changes are caused by your body starting to make and release sex hormones. For males the sex hormone is called testosterone. Females have two hormones that control these changes: oestrogen and progesterone.

Testosterone

For boys, puberty starts when the testes start to make and release testosterone. Testosterone triggers many of the different physical changes that boys experience during puberty. These include:

- growing taller and stronger
- deepening voice
- penis growth
- sperm production
- growing facial hair and pubic hair.

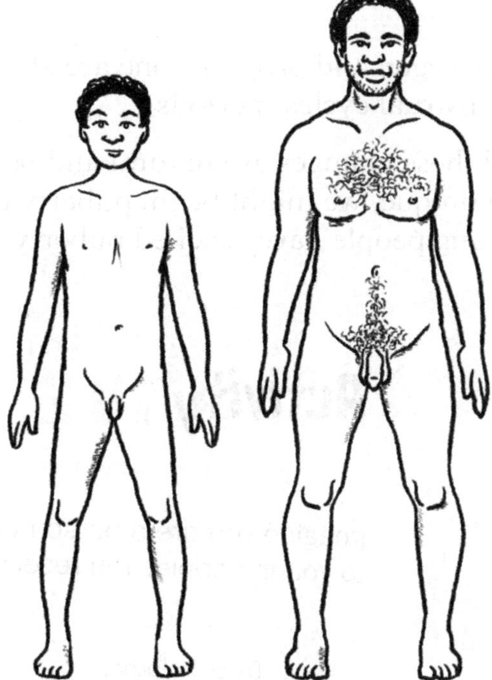

Oestrogen and progesterone

For girls, puberty starts when the ovaries start to produce oestrogen and progesterone. Oestrogen in the bloodstream triggers many of the different physical changes that girls experience during puberty. These include:

- growing taller
- pubic hair
- breast growth
- hips widening
- menstruation (periods) start releasing one egg a month.

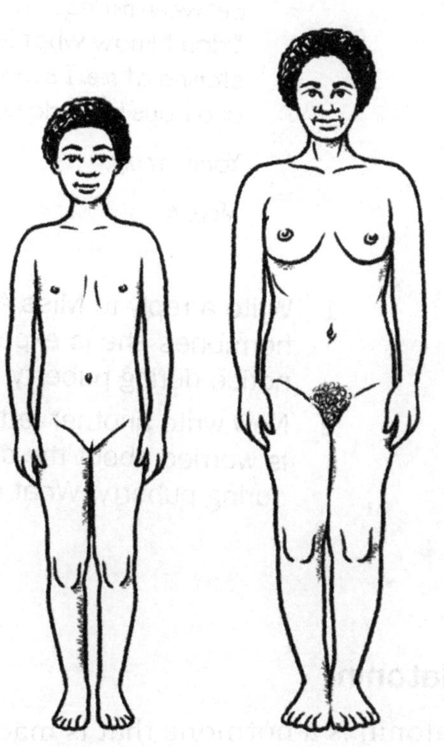

Oestrogen and progesterone are also the two hormones that control girls' menstrual cycles (periods).

All these changes are natural and happen because of hormones. As we are all unique, we might begin puberty before or after our friends, but most young people have reached puberty by the age of 16.

Activity 10.2 PUBERTY HELPLINE

Imagine you are a person who works at a helpline giving advice to young people. You receive this letter:

> Dear Helpline,
>
> I am really worried. Recently I have started to grow hair between my legs and my breasts are getting larger and sore. I don't know what is happening to me and I think everyone is staring at me. I seem to be so much taller than the other girls and boys in my class. Help! What is happening to me?
>
> Yours truly,
>
> Miss A

1 Write a reply to Miss A to help her stop worrying. Talk about the hormones she is experiencing and other changes she might notice during puberty.

2 Now write another letter asking for advice from a boy who is worried about the changes he is feeling in his body during puberty. What questions might he ask?

Melatonin

Melatonin is a hormone that is made by the pineal gland inside your brain. Melatonin is the hormone that tells your body when it is time to go to sleep and when it is time to wake up. It acts as your body clock.

The levels of melatonin in your body change throughout the day. When it is time for you to go to bed, there are higher levels of melatonin in your bloodstream. They lower your body temperature and cause you to feel sleepy.

Activity 10.3 HEALTHY SLEEP

Sleep is very important for children and young people. While you are sleeping your brain develops and your body repairs itself. However, it is sometimes very difficult to get a good night's sleep, especially if you live in a noisy place or in a crowded dormitory.

1 Investigate how many hours of sleep different people get each day and night.

Person	1st night	2nd night	3rd night	Average
Baby				
Five-year-old child				
Ten-year-old child				
Fifteen-year-old child				
Adult				
Old person				

Adults should be getting at least eight hours of sleep a night. Children should get a lot more.

2 What do your results tell you?

3 List some problems caused by lack of sleep.

4 List some ways of getting a good night's sleep. For example, avoiding coffee, tea and cola before sleeping, sleeping in a dark room, using a mosquito net, etc.

The most important gland in the body – the boss!

The pituitary gland is the most important endocrine gland in your body. It is located in your brain and is only about the size of a pea! Your pituitary gland makes eleven different kinds of hormones and controls five other glands around the body. It is the master gland.

The pituitary gland controls:

- how tall you will be
- when you start puberty
- milk production in breastfeeding
- how fast your body systems function.

Changing hormone levels

Doctors and scientists have learned to use hormones to change the way a person's body functions. An example of this is in family planning methods such as birth control pills and injectible contraceptives (e.g. Depo Provera).

With injectible contraceptives, a health worker gives an injection to a woman every three months. The injection includes the female hormone progesterone. High levels of progesterone prevent ovulation (release of an egg from a woman's ovary) so that she cannot become pregnant.

With birth control pills, a woman takes a pill each day. The pills include the female hormones oestrogen and progesterone. High levels of these hormones prevent ovulation so that a woman cannot become pregnant.

Both of these family planning methods are available in PNG. They are easy and safe to use. Used properly, they are very effective at preventing pregnancy. When a woman stops using them she can become pregnant as normal, although this may take a couple of months if she has had the injection.

Besides preventing pregnancy, some of the other benefits of these family planning methods include:

- shorter periods
- lighter periods (not as much blood flow)
- fewer cramps with periods
- less tiredness and moods with periods
- less risk of some kinds of cancer (uterus and ovaries).

There can be side effects from using these two family planning methods (for example, blood clots and high blood pressure), so talk to your health worker before using them. Women who smoke, are overweight, have diabetes or are over 35 are more at risk of side effects.

Activity 10.4 LEARNING MORE ABOUT FAMILY PLANNING METHODS: RESEARCH AND INVESTIGATION

Ask your health worker!

Arrange for a health worker to visit your class. Ask them to give a presentation about different family planning methods and how they work. Ask them to show you examples of different family planning methods and explain which methods are the most effective. You can also ask which methods are the most widely used in your area.

Ask your sisters, mothers and friends!

Talk to your sisters, mothers and friends about different family planning methods.

- Has anyone tried contraceptive injections or pills?
- What did they think?
- Were they easy to use?
- Why did they choose that method?

Hormone highs and hormone lows

Too much or too little of any hormone can affect the way your body functions and develops. In some cases, too much or too little of a particular hormone can be harmful.

For example, your thyroid gland plays a very important role in your growth and development from a child to an adult. It makes hormones that help to control the rate of many of your body's functions.

The thyroid gland needs a chemical called iodine to function properly. Usually the body gets enough iodine in the foods and water we eat and drink.

The thyroid can become larger than normal. You can feel it as a lump under your skin at the front of your neck. This lump is often called a **goitre**. If a person has a goitre, it is a sign that they are not getting enough iodine in their diet.

If levels of thyroid hormones are very high your body will use up energy more quickly than normal. Some signs of high thyroid levels include:

- feeling nervous and irritated
- sweating more than normal
- feeling tired but having difficulty sleeping
- fast heartbeat
- weak muscles
- weight loss
- irregular periods (girls)
- skin around eyes becoming red and swollen and eyes appearing to bulge.

If levels of thyroid hormones are too low, this can slow body functions. Some signs of low thyroid levels include:

- tiredness
- slow heart rate
- dry skin
- weight gain.

Activity 10.5 SALT SURVEY

To prevent goitre, iodine is added to many foods that we eat on a regular basis. Most commonly, iodine is added to salt to make sure that everyone gets the iodine they need to keep them healthy.

1 Look at the salt package in your house or at your school. Do you see the word 'iodised' on the package?
2 Do all the different kinds of salt have 'iodised' on their packages?

Salt lick

Do a taste test with your friends and classmates. You will need:

- a packet of iodised salt
- a packet of non-iodised salt
- a blindfold.

1 Ask the person doing the taste test to put on the blindfold.
2 Ask them to taste the two different kinds of salt.
3 Ask them to tell you which salt is iodised. Were they correct?
4 Ask ten people to do the taste test.
5 How many people could tell the difference between iodised and non-iodised salt? What does that tell you?

If you need more information about body systems

 Websites with information about body systems

www.kidshealth.org

www.yucky.discovery.com

www.childrensuniversity.manchester.ac.uk

www.bbc.co.uk/science/humanbody/

Other sources of information

- Your local health worker or doctor
- Trained village health volunteers and trained peer educators
- Teachers

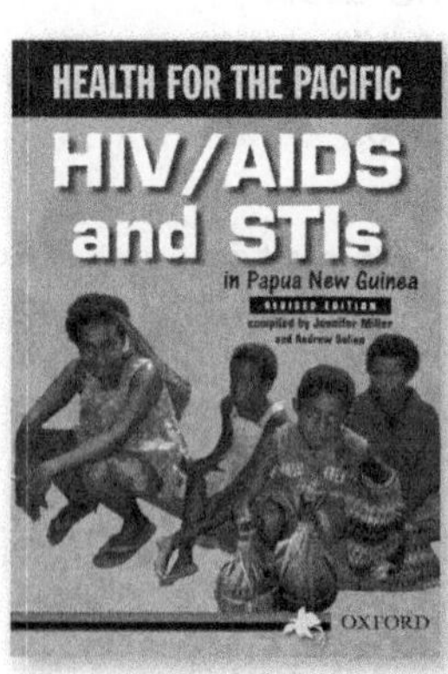

Glossary

alveoli
tiny bags in the lungs where gases are exchanged between the air and the blood

anaemia
a decrease in the number of red blood cells in the blood, causing tiredness and weakness

antibodies
particles produced by the body's immune system in response to an infection

appendix
a small pouch that is connected to your large intestine that no longer has a function in human beings

arteries
blood vessels that carry blood away from the heart through the body

atrium
an upper chamber of the heart. The heart has two atria. Blood enters the heart through the atria.

bacteria
tiny organisms that cause disease and illness

blood transfusion
when health workers give a patient blood from another person. The blood has to be the right blood group.

blood vessels
the collection of tubes (arteries, veins and capillaries) that carry blood throughout your body

bone marrow
flexible, spongy tissue that is found on the inside of bones. In adults it produces new blood cells.

cartilage
smooth, rubbery tissue that connects one bone to another bone. It also pads the ends of bones to prevent them from rubbing against one another

cells
the basic units of a person's body. There are many different types of cells.

communicable diseases
diseases that are caused by a bacteria, virus, parasite or bacteria that you can catch from another human or animal

dehydrated
when the body has lost more liquid than it takes in, causing a person to feel tired, thirsty and dizzy

DNA
the chemical found in all living things that contains the information on how to make that living thing. The chemical is called **deoxyribonucleic acid**.

diaphragm
the layer of muscle below your lungs that helps you to breathe

fertilisation
when a male's sperm and a female's egg join to make a baby

genes
a section of DNA which has a specific function in the body

germ
a micro-organism (for example, a bacterium, virus, or parasite), especially one that causes disease

glands
small organs which make hormones and other fluids

goitre
a swollen thyroid gland usually caused by a lack of iodine

heart attack
a sudden stop of heart function, usually when blood or oxygen supply to the heart is blocked

hormones
chemical messengers

involuntary muscle
a muscle that works without you having to think about it

ligaments
strong cords of tissue that connect bones at moving joints

lymph
fluid containing white blood cells

nerves
special cells for sending electrical signals around the body

organs
collections of cells and tissues that work together to perform a specific function

parasite
a kind of germ that is usually a tiny worm or animal that lives in humans and animals and can cause disease and illness

platelets
small cell pieces that travel through the blood and can stick together to stop bleeding if you have a cut

puberty
the period of time when a child matures and develops into an adult (physically, sexually and mentally)

pulse
the number of times your heart beats in a minute

red blood cells
cells of the blood that carry oxygen to all of your tissues

skeleton
the framework of bones that gives your body structure and shape and protects its important organs

stroke
a sudden stop of brain function, usually when blood or oxygen supply to the brain is blocked

tendons
strong cords of tissue that connect muscles to bones

vaccine
a medicine that gives you a small amount of a particular disease so that your body can make antibodies and protect you if you are ever infected with that disease

veins
blood vessels that carry blood from the body to the heart

ventricle
the lower chamber of the heart. The heart has two ventricles. Blood leaves the heart through the ventricles.

virus
extremely small germs that can cause many infections

voluntary muscle
a muscle that you control with your brain

white blood cells
cells of the immune system that produce antibodies to fight infections